Fate of Microplastics in Wastewater Treatment Plants

This book covers the various sources, the role of treatment technologies, system-associated factors, and future challenges with reference to microplastics in wastewater treatment plants. It also introduces microplastics, their sources, governing factors, microbial diversity effects, and possible control approaches to minimize the exposure of microplastics to human beings. Modelling and distribution of microplastics, environmental sinks, bioindicators, and microplastics as vector in wastewater treatment units are also discussed.

- Focuses on microplastic pollution, mechanism of removal, treatment technologies, pathways, and fate in wastewater treatment system
- Discusses the factors linked to dispersion, survival, and removal efficiency of microplastics in wastewater treatment systems
- Helps understand 'microplastics removal'-centric sustainability aspects of wastewater treatment systems
- Explores the fate of microplastics in sludge-handling systems
- Incorporates comparative case studies from developed and developing nations

This book is aimed at graduate students and researchers in environmental science and engineering, water resources management, wastewater, and chemical engineering.

Fate of Microplastics in Wastewater Treatment Plants

Occurrence, Identification, Potential Factors, and Future Perspectives

Edited by

Nitin Kumar Singh, Komal Jayaswal, Manish Yadav, and Namita Maharjan

CRC Press
Taylor & Francis Group
Boca Raton London New York

CRC Press is an imprint of the
Taylor & Francis Group, an **informa** business

Designed cover image: Shutterstock

First edition published 2025
by CRC Press
2385 NW Executive Center Drive, Suite 320, Boca Raton FL 33431

and by CRC Press
4 Park Square, Milton Park, Abingdon, Oxon, OX14 4RN

CRC Press is an imprint of Taylor & Francis Group, LLC

ISBN: 978-1-032-53877-8 (hbk)
ISBN: 978-1-032-57307-6 (pbk)
ISBN: 978-1-003-43879-3 (ebk)

DOI: 10.1201/9781003438793

Typeset in Times
by codeMantra

Contents

About the Editors

Dr Nitin Kumar Singh is presently an Associate Professor in the Department of Chemical Engineering at Marwadi University, Gujarat, India. He earned his PhD from the Indian Institute of Technology Roorkee in Environmental Engineering, MTech from the Indian Institute of Technology Kharagpur in Environmental Engineering, and BTech from the Institute of Engineering and Technology Lucknow in Chemical Engineering. He has more than 12 years of experience in the field of wastewater treatment, waste management, green synthesis and processes, cost-benefit analysis, biological air pollution monitoring, and sustainability assessment of environmental systems. As of now, he has published over 39 SCI articles, 16 book chapters, and 12 papers in international conferences and edited two international books with international publisher.

Manish Yadav is presently Manager at Central Mine Planning Design Institute Limited, a subsidiary of Coal India Limited, India. He did his Master's degree in Environmental Engineering and BTech in Biotechnology. He has more than 10 years of experience in the mining industry, especially in the environment division. He is NABET-approved expert in Air Pollution (source apportionment study, air quality modelling, air quality assessment) and Water Pollution (wastewater treatment, adsorption, groundwater quality assessment). His core area of expertise is Environmental Impact Assessment. He has published more than 17 papers in reputed international journals along with 12 book chapters in reference books. He has also prepared many consultancies to report on environmental problems. He is appreciably skilled in various data analysis and modelling software including AERMOD, Aquachem, Openair, R software, SPSS, Origin, ArcGIS, ArcSWAT, and ERDAS.

Dr. Komal Jayaswal is currently an Assistant Professor in the Department of Civil Engineering at GITAM University, Hyderabad. She completed her PhD in Environmental Engineering from IIT Roorkee. She did her Master's degree in Environmental Engineering at NIT Allahabad and her Bachelor's degree in Biotechnology at SHIATS Allahabad. Her area of research includes energy-efficient and nature-based technology for wastewater and water treatment system, anaerobic wastewater treatment system, sludge characterization, and solid waste management. She has 10 years of experience in environmental waste management and wastewater treatment and has worked for both field-scale and lab-scale wastewater treatment technology. She has published five research articles and three book chapters and presented her work at international conferences.

Dr. Namita Maharjan is presently an Assistant Professor in the Department of General Education at the National Institute of Technology, Nagaoka, Niigata, Japan. She did her PhD and MTech in Environmental Engineering from the National Institute of Technology, Nagaoka, Japan. Her areas of research include the assessment of wastewater treatment system, microbiology of wastewater treatment plants, and aerobic post treatment system. She has more than 10 years of experience in the field of wastewater treatment systems with expertise at laboratory and field scales. She has published more than 15 research articles and book chapters, and has presented her work at several international conferences.

Contributors

Mohamed Alaraby
Zoology Department, Faculty of Science
Sohag University
Sohag Al Gadida City, Egypt

Muhammad Ali
Institute of Agro-Industry and Environment, Faculty of Agriculture and Environment
The Islamia University of Bahawalpur
Bahawalpur, Pakistan

Kassian T. T. Amesho
Institute of Environmental Engineering
National Sun Yat-Sen University
Kaohsiung City, Taiwan

Muhammad Anas Khan
Jianshui Research Station, School of Soil and Water Conservation
Beijing Forestry University
Beijing, China

Seren Acarer Arat
Environmental Engineering Department
İstanbul University-Cerrahpaşa
Avcılar, Turkey

Muhammad Arslan Aslam
Faculty of Science and Technology
Research Centre for Experimental Marine Biology and Biotechnology
Bilbao, Spain
and
Department of Energy, Environment and Climate Change
Asian Institute of Technology (AIT)
Pathum Thani, Thailand

Ali Ashry
Zoology Department, Faculty of Science
Sohag University
Sohag, Egypt

Zohre Avakh
NAT-Co (Newsad Ab Tarava Co.)
AUT Technology Tower (Ibn Sina)
Tehran, Iran

Manish Chaudhary
School of Environment & Natural Resources
Doon University
Uttarakhand, India

Chingakham Chinglenthoiba
Department of Chemistry
National University of Singapore
Singapore, Singapore
and
Centre of Research Impact and Outcome
Chitkara University
Rajpura- 140417 Punjab, India.

Mary Carolin Kurisingal Cleetus
Faculty of Science and Technology
Research Centre for Experimental Marine Biology and Biotechnology
Bilbao, Spain
and
Faculty of Sciences
University of Liege
Liege, Belgium

D. A. N. D. Daranagama
Department of Technology, Faculty of Indigenous Health Sciences and Technology
Gampaha Wickramarachchi University of Indigenous Medicine
Yakkala, Sri Lanka

Dorcas Akua Essel
Faculty of Science and Technology
Research Centre for Experimental Marine Biology and Biotechnology
Bilbao, Spain
and
Department of Climate Change Impacts in Oceans and Coast
AZTI Foundation: Marine Research Center
Sukarrieta, Spain

Asadulla Hil Galib
Faculty of science and Technology
University of Bordeaux
Bordeaux, France
and
Faculty of Science and Technology
University of Bordeaux
Talence, France

Revantkumar S. Gorfad
Department of Chemistry
Marwadi University
Rajkot, India

Azfar Hussain
International Research Center on Karst under the auspices of UNESCO
Institute of Karst Geology
Chinese Academy of Geological Sciences
Beijing, China

Sioni Iikela
The International University of Management
Centre for Environmental Studies
Windhoek, Namibia

Muhammad Irfan
Institute of Agro-Industry and Environment, Faculty of Agriculture and Environment
The Islamia University of Bahawalpur
Bahawalpur, Pakistan

Mahdi Javanmardi
School of Materials and Advanced Processes Engineering, Department of Textile Engineering
Amirkabir University of Technology
Tehran, Iran

Timoteus Kadhila
School of Education, Department of Higher Education and Lifelong Learning
University of Namibia
Windhoek, Namibia

Mohammad Karimi
School of Materials and Advanced Processes Engineering, Department of Textile Engineering
Amirkabir University of Technology
Tehran, Iran

Azza M. Khedre
Zoology Department, Faculty of Science
Sohag University
Sohag, Egypt

Haleh Khoramshahi
School of Materials and Advanced Processes Engineering, Department of Textile Engineering
Amirkabir University of Technology
Tehran, Iran

Mohd Nizam Lani
Microplastic Research Interest Group
Universiti Malaysia Terengganu
Kuala Terengganu, Malaysia

D. L. C. P. Liyanage
Department of Information and Communication Technology, Faculty of Technology
University of Sri Jayewardenepura
Nugegoda, Sri Lanka

Anik Majumdar
Division of Plant Pathology
ICAR-Indian Agricultural Research Institute
New Delhi, India

Kavitha Malarvizhi
TNJFU-Institute of Fisheries Post Graduate Studies
Chennai, India

Sorour Ayoubian Markazi
School of Materials and Advanced Processes Engineering, Department of Textile Engineering
Amirkabir University of Technology
Tehran, Iran

Suranjana V. Mayani
Department of Chemistry
Marwadi University
Rajkot, India

Chandra Mohan
Department of Chemistry, School of Basic and Applied Science
K R Mangalam University
Sohna, India

K. Murugesan
R&D wing, CMC
State Pollution Control Board
Bhubaneswar, India

Parsa Mushtaq
Research Center for Urban Forestry
Beijing Forestry Unisversity
Beijing, China

Komal Rani Narejo
School of Computer and Artificial Intelligence
Zhengzhou University
Zhengzhou, China

B. S. Panda
R&D Wing, CMC
State Pollution Control Board
Bhubaneswar, India

Ashutosh Pandey
Institute for Water and Wastewater Technology
Durban University of Technology
Durban, South Africa

Kalaiselvan Pandi
TNJFU-Institute of Fisheries Post Graduate Studies
Chennai, India

Meet R. Parmar
Department of Chemistry
Marwadi University
Rajkot, India

S. S. Pati
R&D Wing, CMC
State Pollution Control Board
Bhubaneswar, India

Prabhakaran M. P.
Department of Aquatic Environment and Management
Kerala University of Fisheries and Ocean Studies
Kochi, India

Somaia A. Ramadan
Zoology Department, Faculty of Science
Sohag University
Sohag, Egypt

Amit Ranjan
TNJFU-Institute of Fisheries Post Graduate Studies
Chennai, India
and
J. Jayalalithaa Fisheries University
Nagapattinam, India

Vinitkumar K. Rathod
Department of Chemistry
Marwadi University
Rajkot, India

Taqi Raza
Department of Biosystems Engineering & Soil Science
University of Tennessee
Knoxville, Tennessee

Ansa Rebi
Jianshui Research Station, School of Soil and Water Conservation
Beijing Forestry University
Beijing, China

Rejish Kumar V. J.
Faculty of Ocean Science and Technology
Kerala University of Fisheries and Ocean Studies
Kochi, India
and
Department of Aquaculture
Kerala University of Fisheries and Ocean Studies
Kochi, India

Herish N. Ribadiya
Department of Chemistry
Marwadi University
Rajkot, India

A. Sahu
P.G. Department of Zoology
Utkal University
Vani Vihar, India

S. K. Sahu
P.G. Department of Environmental Sciences
Sambalpur University
Jyoti Vihar, India

Ajay Valiyaveettil Salimkumar
Faculty of Science and Technology
Research Centre for Experimental Marine Biology and Biotechnology (PIE-UPV/EHU)
Plentzia, Spain

R. K. Sethi
P.G. Department of Environmental Sciences
Sambalpur University
Jyoti Vihar, India

Sumarlin Shangdiar
Institute of Environmental Engineering
National Sun Yat-Sen University
Kaohsiung, Taiwan

Devansh Sharma
Department of mechanical engineering
University of British Columbia
Kelowna, Canada

Nallin Sharma
Materials Engineering Department
Indian Institute of Sciences
Bengaluru, India

Dharmesh H. Sur
Department of Chemical Engineering, Faculty of Technology
Marwadi University
Rajkot, India

Surindra Suthar
School of Environment & Natural Resources
Doon University
Dehradun, India

Masoumeh Sharifi Teshnizi
School of Materials and Advanced Processes Engineering, Department of Textile Engineering
Amirkabir University of Technology
Tehran, Iran

Divya O. Tirva
Department of Chemical Engineering, Faculty of Technology
Marwadi University
Rajkot, India

Guan Wang
Jianshui Research Station, School of Soil and Water Conservation
Beijing Forestry University
Beijing, China

M. Yadav
Central Mine Planning and Design Institute, RI-VII
Bhubaneswar, India

Jinxing zhou
Jianshui Research Station, School of Soil and Water Conservation
Beijing Forestry University
Beijing, China

Preface

The ubiquitous presence of microplastics, tiny plastic fragments, has become a pressing environmental concern in recent years. These persistent pollutants contaminate our water resources, raising significant questions about their impact on human health and ecological systems. Wastewater treatment plants are identified as a major pathway for microplastics to enter the environment, acting as both sources and sinks for these microscopic contaminants. With these concerns, this book delves deep into this complex and multifaceted issue. It offers a comprehensive exploration of the presence, behaviour, and removal of microplastics within the wastewater treatment context. This book is designed to serve as a valuable resource for a wide audience, including researchers and academics in environmental science, wastewater treatment, and microplastics research, engineers and practitioners working on wastewater treatment technologies, policymakers, students, and individuals interested in understanding the fate of microplastics in wastewater treatment. This book particularly focuses on key areas such as the presence and characterization of microplastics, health and environmental impacts, bioindicators, removal approaches for microplastics, environmental sinks, and the applications of artificial intelligence and machine learning tools for microplastics detection and/or characterization.

The issue of microplastics in wastewater presents a significant challenge, but also an opportunity for innovation and collaboration. This book aims to contribute to ongoing efforts by providing a comprehensive overview of the current state of knowledge and highlighting promising avenues for future research and action. By working together, we can develop effective solutions to address this challenge and ensure a cleaner and healthier environment for all. The team of editors invites all interested readers to embark on this journey with us, exploring the complex world of microplastics in wastewater and seeking new pathways towards a sustainable future.

Nitin Kumar Singh

Komal Jayaswal

Manish Yadav

Namita Maharjan

Acknowledgements

We sincerely wish to thank Dr Gagandeep Singh, Swapnil Joshi, Jahnavi Vaid, and Aditi Mittal from CRC Press for their excellent guidance, support, and coordination of this fascinating project. We would like to thank all the contributors for sharing their knowledge and experiences in the form of contributed chapters. We greatly appreciate the support from all the reviewers for spending their valuable time reviewing the chapters of this book. We are also thankful to our family members, friends, and other academic colleagues who showed patience when we deprived them of our much-needed attention during the course of the book assignment. We would like to acknowledge the support received from the management and competent authority of Marwadi University (India), GITAM University (India), Central Mine Planning and Design Institute (India), and the National Institute of Technology (Japan). Last but not least, we are also thankful to Almighty God who has awarded us the wisdom to explore such environmental issues and complete this book. We believe that despite our dedication, the first version always comes with some errors that may have crept into this compilation. We are happy to receive constructive feedback from readers to improve our future volumes.

Introduction

Microplastics, which are tiny fragments and fibres measuring less than 5 mm in size, have infiltrated our wastewater treatment systems, posing a silent yet significant threat to human health and ecological well-being. This book delves into the heart of this complex issue and helps readers understand the processes by which these small invaders navigate wastewater treatment plants, as well as the various factors that influence their fate and behaviour. This book is not just a scientific exploration; it is also a call to action for the management of microplastics in wastewater treatment systems. The editors have summarized information covering a wide range of topics, including the characteristics, quantification, and distribution of microplastics, environmental sinks, bioindicators, and microplastics as vectors in wastewater treatment units. By understanding the presence, threats, detection approaches, and potential solutions for removing microplastics within wastewater treatment systems, we aim to collectively work towards a cleaner, healthier future for both ourselves and the environment. The journey of this book begins with the establishment of a scientific foundation and an understanding of the research trends in the context of wastewater treatment, followed by the characterization and quantification aspects of microplastics. Subsequent chapters focus on solutions, case studies, and advancements in the characterization, detection, and monitoring of microplastics using artificial intelligence and machine learning.

The first chapter focuses on the analysis of the microplastics problem through bibliometric analysis, uncovering the extensive research landscape surrounding microplastics in wastewater. Chapters 2–4 highlight the characterization techniques for microplastics and their threats to human health and the environment. Additionally, bioindicators, nature's silent sentinels, are introduced as tools for monitoring microplastic contamination, demonstrating the universality of this issue. In Chapters 5–7, the roles of various microbes and their enzymatic mechanisms, impacts on aquaculture, and investigation approaches in wastewater treatment systems are critically reviewed and presented. Chapter 8 explores the characterization and removal of microplastics at different stages of the treatment process, identifying potential hotspots and inefficiencies. Meanwhile, Chapter 9 highlights the concept of the environmental sink associated with wastewater treatment, emphasizing the need for holistic solutions. In Chapter 10, mass balance and life cycle aspects are also critically discussed for microplastics. Chapters 10 and 12 emphasize the importance of local action by showcasing the unique challenges and potential solutions specific to different contexts. The last two chapters of this book explore the cutting-edge field of machine learning and artificial intelligence in microplastic detection.

Overall, this book highlights that through collaborative efforts and innovative solutions, we can strive towards a cleaner future where our wastewater systems are not conduits for microplastic pollution, but rather effective barriers that protect our environment and our health.

1 Examining the Presence of Microplastics in Wastewater

A Bibliometric Analysis and Overview

S.S. Pati, B.S. Panda, R.K. Sethi, A. Sahu, M. Yadav, K. Murugesan, and S.K. Sahu

1.1 INTRODUCTION

Plastics are widely used for various applications in various sectors including personal care, apparel, medical and industrial sectors. This is attributed to their lightweight, stable chemical properties, corrosion resistance, and affordability (Hernandez et al., 2017). According to a forecast, global plastics production was expected to reach 390.7 million tons in 2021, representing an annual growth rate of 4% (GPP, 2023). The escalation of manufacturing and use has led to an unprecedented proliferation of plastic waste and widespread contamination. Most plastics used by consumers are single-use and have limited recycling. Globally, only 9% of plastic waste has been recycled, 12% has been incinerated, and the remaining 79% remains in various ecosystems (Geyer et al., 2017). According to a study (Borrelle et al., 2020), it is expected that around 53 million tons of plastic waste will end up in water bodies by 2030.

Microplastics (MPs) are defined as plastic polymer particles that measure less than 5 mm. The European Commission has recognized MPs as a significant parameter for monitoring and evaluating seawater pollution that arises from plastic debris. The microplastics (MPs) can be classified into two distinct categories, namely primary and secondary microplastics (Auta et al., 2017). Primary microplastics are generated through manufacturing and packaging procedures, and may also be obtained from cosmetic products that serve as exfoliants or pharmaceutical drugs (Mintenig et al., 2017). Over a period of time, plastic waste of larger dimensions found in various environments such as terrestrial, freshwater, and marine undergo a process of degradation, leading to their fragmentation into smaller particles that are less than 5 mm in size (Horton et al., 2017). The microplastics that are most frequently encountered are sourced from diverse plastic materials, such as polyethylene (PE), polypropylene (PP), polyamide (PA), polyvinyl chloride (PVC), polystyrene (PS), and polyethylene

DOI: 10.1201/9781003438793-1

terephthalate (PET) (Alimi et al., 2018). The emergence of MPs as a significant contaminant has garnered considerable public concern due to its ecological and health impacts, making plastic pollution one of the most critical environmental health issues worldwide (Chae & An, 2017). The extensive use of plastic products globally contributes to the continuous increase in their production each year (Sher et al., 2021).

During the peak period of the COVID-19 pandemic in 2020–2021, the World Health Organization (WHO) recommended the use of face masks to the public and later made it mandatory to minimize the risk of virus transmission (Bhangare et al., 2023). These masks, which are often composed of polymers such as polyethylene (PE), polypropylene (PP), polystyrene (PS), polyurethane (PU), or polycarbonate (PC), have the potential to release microplastic fibers into the environment (Oginni, 2022). Notably, polypropylene, the predominant material used in mask production, degrades rapidly, resulting in the formation of microscopic plastics. However, improper disposal of these masks has become a significant environmental concern, particularly in marine ecosystems, where large quantities of masks have been released from various sources. Ocean Asia reported in 2020 that an estimated one billion masks were discarded into the oceans, basing on an annual production of 52 billion disposable masks. (Xu & Ren, 2021).

Most of these polymers are resistant to degradation by living organisms, necessitating the development of biodegradable substitutes. Existing biodegradable plastics consist of cellulose, polybutylene adipate-co-terephthalate, polybutylene sucrose-co-adipate, po-hydroxyalkanoates, polybutylene succinate, and polylactic acid. However, these plastics often show insufficient decomposition and require a longer decomposition time, only achieving about 90% decomposition under soil conditions (Renzi et al., 2019).

Numerous research investigations worldwide have been dedicated to the study of micro- and nanoplastics. Previous studies have primarily focused on examining various environmental matrices and exploring the fate of microplastics in the environment (Akinpelu & Nchu, 2022). For instance, Papadimitriu and Allinson conducted a study in 2022 to assess the presence of microplastics in marine ecosystems of the Mediterranean Sea. Similarly, Zhou et al. (2022) conducted a comprehensive investigation to evaluate the occurrence of microplastics and nanoplastics in diverse marine environments, encompassing seas, oceans, beaches, bays, gulfs, estuaries, coastlines, and shorelines (Zhou et al., 2022).

The prevalence of microplastics, particularly in the form of primary microplastics, in wastewater is of significant concern. Notably, common food items such as salt, sugar, honey, beer, bottled water, tap water, and fish have been identified as potential sources of plastic contamination, thereby establishing a pathway for microplastic ingestion by individuals (Chang et al., 2020; Zhang et al., 2019). It has been established in previous research that the consumption of contaminated water is a significant route through which microplastics are transmitted from environmental sources to humans (Walker & Fequet, 2023). In a study (Cox et al., 2019), it was revealed that the consumption of food and beverages by an individual could lead to the intake of an estimated 39,000–52,000 numbers of microplastics per year, with variations based on gender and age. Additionally, if bottled water is consumed, this estimate could increase by approximately 90,000 microplastics annually and an additional 4,000 microplastics if tap water is ingested. Owing to their minuscule size, microplastics have the propensity to readily infiltrate hydrological systems and subsequently become integrated into the food chain. However, our current understanding of microplastics remains limited, necessitating urgent research endeavors in this field.

At present, bibliometric analysis is widely acknowledged as an alternative approach for assessing scholarly subjects within the domain of library and information science. It has emerged as one of the most popular methodologies for evaluating and predicting research trends in specific areas, as well as for mapping published records that have been identified (Liu et al., 2021). In addition to providing researchers with enhanced access to a broader range of literature, bibliometric analysis fosters collaboration among researchers by bringing them together on a unified platform (Wagner et al., 2015). These inherent advantages underscore the crucial role of bibliometric analysis as a preliminary step in conducting research on topics of regional, national, or international significance. Several researchers have already conducted bibliometric analyses on various subjects, yielding valuable insights. In the present study, we employ a combined approach of bibliometrics and altmetrics to comprehensively analyze microplastics, offering references and recommendations to future researchers and enabling a deeper understanding of the global trends related to microplastics in wastewater. By employing keyword clusters and the altmetric attention score, we identify the research focal points and potential future directions pertaining to microplastics. Additionally, we conduct a systematic review and summary of the contamination status and challenges associated with microplastics, while also providing suggestions for the future focus and perspectives of microplastics research.

1.2 MATERIALS AND METHODS

1.2.1 Data Collection

A systematic and comprehensive literature search was performed to identify pertinent studies focusing on the presence of microplastics in wastewater. Diverse academic databases, including Scopus, Web of Science, PubMed, Lens, and Dimensions, were meticulously queried using appropriate keywords such as "microplastics," "wastewater," "human health," "food chain," "sewage," "effluent," and related terms. Twelve thousand seven hundred and sixty-two articles were retrieved, out of which 6,359 research articles and reviews of various journals of environmental science were evaluated. The time frame for this investigation was from the earliest accessible date i.e., from 1st January 2000, to 15th June 2023, for a period of 23.5 years.

1.2.2 Inclusion and Exclusion Criteria

Studies meeting specific criteria were included in this analysis. Inclusion criteria encompassed investigations that specifically examined the presence of microplastics in wastewater samples and provided comprehensive information regarding its sampling methods, analytical techniques, and reported results. Conversely, studies focusing on other environmental matrices or lacking specific investigations of microplastics in wastewater were excluded from this analysis. Relevant data were meticulously extracted from the selected studies. All the data were downloaded in EndNote, CSV, RefWorks, and RIS format.

1.2.3 Data Analysis Method

To assess the characteristics and trends of the selected documents, a comprehensive bibliometric analysis was conducted, and when feasible, a statistical test (exponential regression) was employed to evaluate significant differences between groups or variables. The data accumulation and analysis were directed in Microsoft Excel 2021, and related graphs were drawn. Bibliometric indicators, including publication output, author, or organizational or country collaborations, citation analysis, and journal distribution, were meticulously examined using appropriate bibliometric software or programming languages VOSviewer (Version 1.6.13). These analyses provided valuable insights into research productivity, collaboration patterns, and the impact of studies within the field of microplastics in wastewater. The cluster analysis by VOSviewer software generates various social network maps, indicating the importance of the size of a node and the thickness of lines (Padilla et al., 2018). Here, the nodes represent the frequency and the lines connecting the nodes symbolize connotations. Different nodes represent various clustering groups (Chang et al., 2022). If the line is thicker, the relationship between nodes will be greater (Gao et al., 2019). Three different types of visualization maps (network, overlay, and density) were derived from VOSviewer and interpreted. The "co-occurrence keywords" is a prominent way to get an idea about the main content of the research field. The extracted data were thoughtfully synthesized to provide a comprehensive overview of the presence of microplastics in wastewater. The synthesized findings were systematically categorized based on geographical locations, sampling methods, analytical techniques employed, and the reported concentrations of microplastics in the environment, food chain, and its impacts on humans. This systematic approach enabled the identification of research gaps, emerging trends, and areas requiring further investigation within the field. The limitations associated with the bibliometric analysis were duly recognized and acknowledged. These limitations included potential biases related to the selection of databases, potential language restrictions, and the exclusion of unpublished or non-peer-reviewed sources. The scope of this analysis was strictly limited to the selected studies, and any inherent limitations or biases within the individual studies were thoughtfully considered during the interpretation of the results.

1.3 RESULTS AND DISCUSSIONS

The present investigation entails a thorough bibliometric analysis aimed at scrutinizing the prevalence of microplastics in wastewater and presenting a synopsis of the existing research in this domain. The objective of the study was to ascertain the primary research patterns, notable authors, impact journals, and noteworthy research clusters linked to microplastics in wastewater. The findings of our analysis indicate a consistent rise in the quantity of literature pertaining to microplastics in wastewater during the last 10 years. The increasing trend indicates that scholars are gaining greater cognizance of the prospective hazards and ecological consequences linked with microplastics present in wastewater.

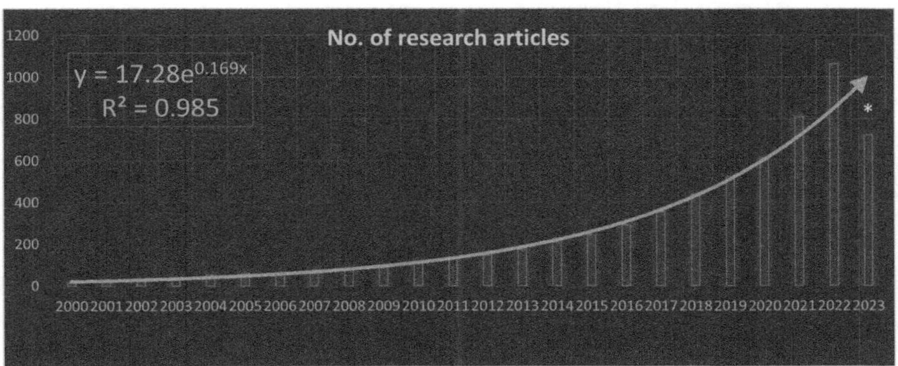

FIGURE 1.1 Literature growth-trend of "microplastic"- and "wastewater"-related research from 2000 to 2023. *Data are for 6 months (from 1st January to 15th June 2023).

1.3.1 Publication Trends and Growth

The findings of our analysis indicate a consistent rise in the quantity of literature pertaining to microplastics in wastewater during the last ten years. The observed expansion (Figure 1.1) denotes an increasing curiosity and acknowledgment of the significance of investigating microplastic pollution in wastewater within academic circles. The increasing trend of published articles from January 1, 2000, to June 15, 2023 ($R^2 = 0.985$) indicates that scholars are gaining greater cognizance of the prospective hazards and ecological consequences linked with microplastics present in wastewater. However, the number of articles published from 2000 to 2010 is almost 642, but will reach 3,121 in the next decade, that is, from 2011 to 2020. The total number of articles published in 2022 on the topic of microplastics and wastewater was 1,062, while in the year 2023, it was 723 within 6 months.

1.3.2 Key Research Topics and Clusters

Utilizing the method of keyword co-occurrence analysis, the present research article aims to shed light on the significant research themes and groupings that emerge within the domain of microplastics in wastewater. By examining the patterns of keyword co-occurrence, a comprehensive understanding of the interrelationships between various research topics can be achieved. In this study, a set of predetermined criteria was employed to select the keywords for analysis. Specifically, keywords with a co-occurrence frequency equal to or greater than 150 were considered, resulting in a pool of 23,046 keywords for evaluation. Notably, only 145 keywords met this threshold, reflecting the selectivity of the analysis and ensuring the inclusion of meaningful and relevant terms. The results of the analysis revealed the presence of distinct research clusters that encapsulate several prominent research themes. These clusters encompass a range of topics, including "wastewater treatment," "microplastic characterization," "microplastic fate and transport," "impact on aquatic ecosystems," and "health implications." The identification of these clusters emphasizes the interdisciplinary nature of microplastic research, highlighting the necessity of fostering collaboration among scholars from diverse fields.

TABLE 1.1
Summary and Analysis of Top Co-occurrence Keywords Based on "Microplastic" and "Wastewater"

Co-occurrence Keywords	Occurrences	Total Link Strength
Microplastic	3,003	55,806
Plastic	2,637	55,627
Water pollutant	2,020	46,217
Plastic waste	2,127	45,073
Environmental monitoring	2,025	40,668
Chemical water pollutants	1,989	45,442
Marine environments	1,762	38,849
Polyethylene	1,328	35,691
Controlled study	1,266	28,487
Sediment	1,021	25,676

To further examine the strength and significance of the identified clusters, a cluster density analysis was conducted based on the acquired data. The density of keyword co-occurrence within the clusters, the term "microplastic," notably, exhibited the highest link strength, occurring 3,003 times and demonstrating a link strength of 55,806. This observation underscores the central role of microplastic-related research within the broader context of the investigated domain. Additionally, Table 1.1 presents the top ten keywords with their corresponding occurrence frequencies and link strengths. These keywords provide insights into the most prominent and influential terms within the analyzed dataset, further contributing to our understanding of the research landscape surrounding microplastics in wastewater. By employing rigorous criteria and analyzing a substantial number of keywords, the study provides a comprehensive overview of the research landscape and offers valuable insights for future investigations in this domain. However, the number of co-occurrences of "treatment method" and "human physiology" is less than 40.

Previous studies on the treatment of "microplastics" and its effects on human physiology have primarily concentrated on the current processes utilized in sewage treatment plants and drinking water treatment facilities (Ma et al., 2019; Liu et al., 2021). Additionally, investigations have been conducted on the potential presence and impact of microplastics within the human body (Schwabl et al., 2019; Vianello et al., 2019). In subsequent times, this approach may gain widespread acceptance, and the associated focal areas may pertain to conventional ecological remediation methodologies like engineered wetlands, biodegradation (involving insects capable of consuming and decomposing plastics), and photocatalytic technology besides plastic human physiology.

1.3.3 Influential Country, Institutions, and Authors

1.3.3.1 Dominant Countries in Microplastics Research

A comprehensive investigation was conducted to identify the countries and institutions that have significantly contributed to the field of microplastics in wastewater research. The study employed a rigorous criterion that required a minimum number of documents from a country, with at least 120 articles and 2,000 citations, to be included in the analysis. Out of the 149 countries that publish journals on microplastics and wastewater, only 20 countries met these established thresholds. China emerged as the most prominent country in this research domain, securing the top rank and contributing to 30% of the global research output. Following China, the United States, the United Kingdom, Germany, and Australia were among the top-ranking countries, all fulfilling the aforementioned criteria. The findings also revealed strong linkages between these countries (Figure 1.2a), indicating collaborative efforts and information exchange within the global research community. Furthermore, cooperation studies are centered between China and South Korea, Japan, the United States, Australia, India, and Hong Kong. The United Kingdom, Canada, Turkey, Germany, the Netherlands, and Norway have rather tight research ties. France also established more collaboration with Spain, Italy, Portugal, and Brazil than with any other country.

Furthermore, the analysis aimed to identify the most notable authors and institutions in the field of microplastics in wastewater. Once again, China demonstrated its predominance in this regard. The same entities mentioned earlier, namely China, the United States, the United Kingdom, Germany, and Australia, were found to have made significant contributions to this realm of research (Figure 1.2b). These contributions have played a crucial role in advancing our understanding of microplastic

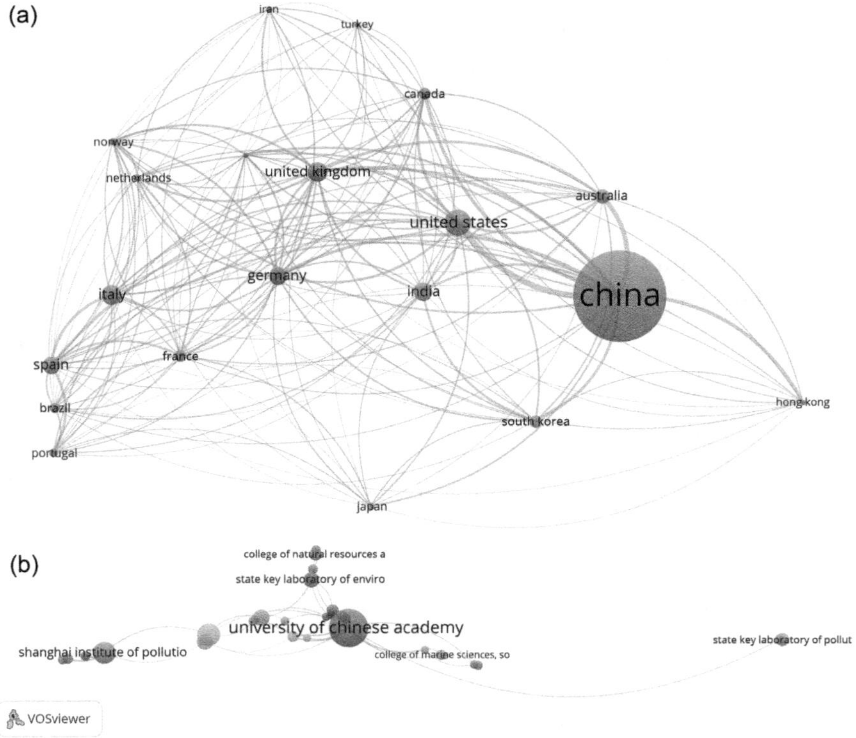

FIGURE 1.2 Network visualization map of top 20 productive countries. (a) Collaboration networks map, (b) cluster analysis map.

pollution in wastewater. The academic achievements of the individuals associated with these entities have had a substantial impact on the trajectory of the field and have influenced subsequent inquiries.

1.3.3.2 Pioneering Institutions in Microplastics Study

In addition to assessing countries and authors, the study also focused on identifying institutions that have actively contributed to the research on microplastics in wastewater (Table 1.2). Using a criterion of a minimum of 20 documents and 100 citations, out of the 12,690 organizations publishing journals on this topic, only 29 institutes met the established thresholds. These institutions represent a select group that has consistently produced substantial research output and garnered significant recognition within the scientific community. The findings of this research article highlight the global landscape of microplastics in wastewater research, emphasizing the prominent role of China, along with other influential countries such as the United States, the United Kingdom, Germany, and Australia. The study also acknowledges the significant contributions made by specific authors and institutions in advancing the knowledge in this critical area of study.

Overlay cooperation networks between nations reveal scientific production and collaboration tendencies. This study examined the average number of publications each year from January 1, 2019, through June 15, 2023, the most active period. The cooperation networks of 33 of the 149 nations with at least 50 papers were examined. Each nation is a node in the overlay collaboration networks, and node connections reflect country partnerships. Countries with more annual publications tend to collaborate more. The size of each node is proportional to the total number of publications during the most prolific era to find nations with at least 50 papers. It identifies the

TABLE 1.2
Top Five Institutes Conducting Research on "Microplastic" and "Wastewater"

Rank	Organization	Number (% Global Contribution)
1	University of Chinese Academy of Sciences, Beijing, China	106 (1.67)
2	State Key Laboratory of Estuarine and Coastal Research, East China Normal University, Shanghai, China	48 (0.75)
3	Shanghai Institute of Pollution Control and Ecological Security, Shanghai, China	46 (0.72)
4	State Key Laboratory of Freshwater Ecology and Biotechnology, Chinese Academy of Sciences, Wuhan, China	30 (0.47)
5	State Key Laboratory of Environmental Criteria and Risk Assessment, Chinese Research Academy of Environmental Sciences, Beijing, China	27 (0.42)

United States, China, the United Kingdom, the Netherlands, South Korea, Australia, and India as important stakeholders with strong collaborative strengths.

1.3.3.3 Cited Research Works and Their Impact

Investigating the global scientific contribution, it is revealed that only 67 out of 149 countries fulfilled the criteria of a minimum of 30 papers and 500 citations per nation (Table 1.4). This indicates that a considerable number of countries have yet to achieve substantial scientific output in terms of paper publication and citations. Interestingly, China, which is the major research contributor, did not surpass the average number of papers per nation. This suggests that China's research output is not significantly higher than that of other countries on average. However, it is worth noting that the University of Chinese Academy of Sciences in Beijing, China, ranked first in publishing research articles, with a total of 106 papers contributing to 1.67% of global research output. In contrast, among the top 15 countries, the Netherlands exhibited the highest citation rate per article, with an impressive average of 114.34 citations per paper, as shown in Table 1.3. This underscores the influential and esteemed nature of the scientific work conducted in the Netherlands. The United Kingdom closely followed with an average citation rate of 98.66, further emphasizing the quality and impact of their research output. These findings highlight the varying levels of scientific productivity and impact across nations, with some countries meeting the minimum thresholds, while others have room for improvement. Citation rates of the Netherlands and the UK demonstrate their scientific achievements. These nations have produced scientifically significant research. However, the study provides useful insights into global scientific contribution and emphasizes the necessity of research and innovation for global scientific understanding.

TABLE 1.3
Summary of Top 15 Country Distribution with Citation Time

Country	No. of Articles	Citations	Average Nos. of Citations per Article	Total Linkage Strength
China	1,915	72,162	37.68	773
United States	483	25,317	52.42	574
United Kingdom	350	34,531	98.66	441
Germany	316	18,928	59.90	365
Australia	247	13,664	55.32	350
India	319	11,296	35.41	335
Spain	314	10,754	34.25	282
Italy	353	13,350	37.82	275
France	196	12,570	64.13	258
Canada	191	11,293	59.13	223
Norway	127	8,940	70.39	218
South Korea	203	8,487	41.81	207
Netherlands	118	13,492	114.34	183
Portugal	147	5,533	37.64	159
Brazil	172	6,023	35.02	156

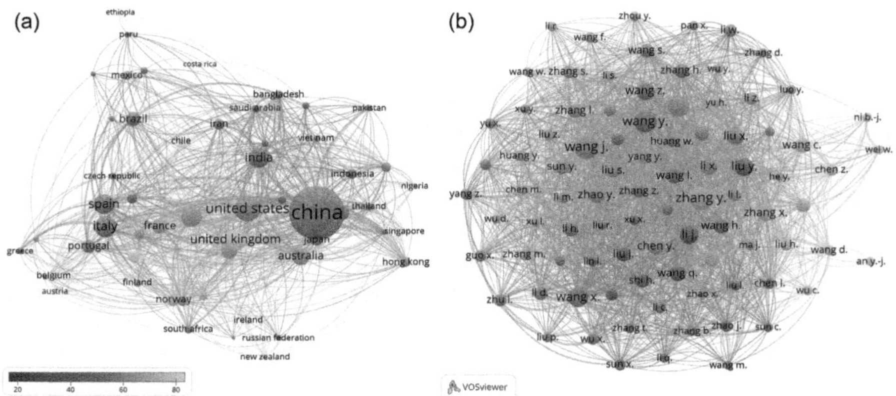

FIGURE 1.3 (a) Overlay collaboration networks among the most productive countries with average citations per year of published articles. (b) Cluster analysis of authors highly cited in the field of microplastic in wastewater during the period of 2000–2023 (till June 15, 2023).

1.3.3.4 Leading Authors and Their Contributions

When assessing authors who have actively contributed to the research on microplastics in wastewater, using a criterion of a minimum of 30 documents and 80 citations, out of the 14,480 authors, only 93 met the established thresholds (Table 1.4 and Figure 1.3). The findings reveal that J. Wang, affiliated with an institution in China, has emerged as the predominant author in terms of the number of publications on the topic under investigation. Table 1.4, a crucial component of this research, displays the distribution of publications among different authors and their corresponding link strengths. Link strength is a measure that indicates the impact and influence of an author's work within the scientific community. Higher link strength signifies a greater citation rate, recognition within the field, and potential to shape future research directions in the field. The dominance of J. Wang, in terms of publications and link strength, underscores their expertise, productivity, and influence in the research area. It also suggests that their work has gathered substantial attention and recognition from peers and scholars worldwide. However, the article entitled published by Hidalgo-Ruz and his research group is the most cited article in the journal *Environmental Science and Technology,* focusing on "microplastics in the marine environment" with citation of 2,734 times (Table 1.5), based on sampling and analysis method (Hidalgo-Ruz et al., 2012). The foremost areas of research which are cited more includes, the examination of microplastics for human consumption (Van Cauwenberghe & Janssen, 2014), the evaluation of the impact of debris on marine life (Gall & Thompson, 2015), the analysis of microplastics in the aquatic environment (Murphy et al., 2016), the identification of the source of microplastics in the environment (Dris et al., 2016), and transportation and ultimate destiny of microplastic in wastewater treatment plants (Liu et al., 2021). The topics also covered the global inventory of small floating plastic debris (Van Sebille et al., 2015), recommendations and categorization framework for plastic debris (Hartmann et al., 2019), microplastics in marine pollution caused by facial cleansers (Fendall & Sewell et al., 2009), and microplastic pollution in lakes (Free et al., 2014).

TABLE 1.4
Top 10 Authors of Various Countries with Their Publications and Citations

Rank	Author	Country	No. of doc.	Total doc of Country	% Country Wise	Total Doc.	% Globally	Total Citations	Total Link Strength
1	Wang, J.	China	148	1,915	7.73	6,359	2.33	8,080	752
2	Carr, S.A.	United States	102	483	21.12	6,359	1.60	6,917	537
3	Van Sebille, E.	United Kingdom	90	350	25.71	6,359	1.42	6,132	413
4	Funck, M.	Germany	86	316	27.22	6,359	1.35	4,630	345
5	Galafassi, S.	Australia	99	247	40.08	6,359	1.56	4,084	330
6	Sharma, S.	India	71	319	22.26	6,359	1.12	2,703	328
7	Paul-Pont, I.	Spain	101	314	32.17	6,359	1.59	3,003	272
8	Avio, C.G.	Italy	82	353	23.23	6,359	1.29	2,529	256
9	Dris, R.	France	73	196	37.24	6,359	1.15	2,898	238
10	Alimi, O.S.	Canada	79	191	41.36	6,359	1.24	3,122	226

TABLE 1.5
Most Cited Articles Published during the Period of 2000–2023 (till June 15, 2023) on "Microplastic" and "Wastewater"

Rank	Article	Journal	Main Research Interest	Citations
1	Hidalgo-Ruz et al. (2012)	*Environmental Science and Technology*	Microplastics in the marine environment	2,734
2	Van Cauwenberghe and Janssen (2014)	*Environmental Pollution*	Microplastics for human consumption	1,185
3	Gall and Thompson (2015)	*Marine Pollution Bulletin*	Impact of debris on marine life	1,170
4	Murphy et al. (2016)	*Environmental Science and Technology*	Microplastics in the aquatic environment	1,066
5	Dris et al. (2016)	*Marine Pollution Bulletin*	A source of microplastics in the environment	1,024
6	Carr et al. (2016)	*Water Research*	Transport and fate of microplastic particles in wastewater treatment plants	988
7	Van Sebille et al. (2015)	*Environmental Research Letters*	Global inventory of small floating plastic debris	951
8	Hartmann et al. (2019)	*Environmental Science and Technology*	Recommendations for and categorization framework for plastic debris	918
9	Fendall and Sewell (2009)	*Marine Pollution Bulletin*	Microplastics in marine pollution by facial cleansers	904
10	Free et al. (2014)	*Marine Pollution Bulletin*	Microplastic pollution in lakes	863

1.3.4 Journals and Publication Outlets

The present study aimed to identify the key journals that have made significant contributions to the field of microplastics in wastewater research. These journals play a pivotal role in disseminating research findings and fostering scientific discourse within this particular domain. Analyzing the publication channels enables scholars to determine the most relevant periodicals for sharing their research while also empowering decision-makers and interested parties to access the most up-to-date information on microplastics in wastewater. Through a comprehensive analysis, it was found that the journal *Marine Pollution Bulletin* emerged as the primary publication outlet for research on microplastics in wastewater, with a notable number of articles published (376). This journal holds an h-index value of 193, indicating a significant impact and influence within the scientific community. The findings suggest that researchers in this field have recognized *Marine Pollution Bulletin* as a reputable platform for disseminating their work and engaging in scientific discussions.

Following the journal *Marine Pollution Bulletin*, two other journals such as *Environmental Pollution* and *Science of the Total Environment* were identified as influential publication venues for research on "microplastics in wastewater." While the number of articles published in these journals was slightly lower than *Marine*

Pollution Bulletin, their significance in advancing knowledge in this area should not be overlooked. Moreover, it is worth noting that *Environmental Science and Technology*, despite publishing a comparatively smaller number of articles (239), demonstrated the highest h-index among the identified journals. This indicates that the research published in this journal has had a substantial impact and has been widely cited within the scientific community. *Environmental Science and Technology* continues to serve as an important resource for researchers and interested stakeholders seeking the latest advancements and insights on "microplastics in wastewater."

This detailed study underscores global concerns about the impact of microplastic contamination on human health through the food chain. The prevalence of microplastics in various food sources and their potential health impacts highlight the urgent need for further studies, risk assessments, and containment techniques. International collaboration and standardized approaches are needed to create effective solutions to reduce microplastic pollution and protect human health. Further research on the toxicity, fate, and transport of microplastics is needed to develop evidence-based laws and policies to address this growing environmental and health concern. To date, very little research has been conducted worldwide on the impact of microplastics on human health in the food chain. Initially, researchers from 53 countries started working in this research area, of which only 13 countries are actively doing research in this area and are now publishing at least five articles per year. China and India are two prolific countries in this research area. M. Wu and W. Huang from Hunan University, China, J.-J. Guo from Jinan University, China, R. Qi, X. Chang, and L. Yang from the University of Chinese Academy of Sciences, China, M. Kumar from CSIR-Indian Institute of Toxicology Research, Lucknow, India (Kumar et al., 2021), A.C. Vivekanand from KTH Royal Institute of Technology, Sweden, in collaboration with M.J. Taherzadeh from the Indian Institute of Technology Roorkee, India are some of the top researchers in this field. There is still a lot of work to be done in this area in the future as it is an emerging field of research.

1.4 CONCLUSION

This bibliometric study shows that microplastics have increased significantly in wastewater studies. The amount of literature on this topic has steadily increased over the period of 2000–2023 ($R^2 = 0.985$), indicating a growing interest for microplastic pollution studies in wastewater. The study also highlights the leading countries and institutions in this research area, as well as the geographical distribution of scientific production. Keywords and concurrent clusters showed the importance of microplastics in the research areas of wastewater. The most frequently studied topics were the characterization of microplastics, origin and routes of contamination, treatment and removal, and their environmental impact. The result shows the complexity of microplastic pollution in wastewater and underlines the need for a comprehensive investigation. The present analysis revealed research gaps that require further investigation. Since the methods for collecting, separating, and assessing microplastics differ, uniform standards could increase comparability and ensure research coherence, consistency, and reliability of the data. Although wastewater research focuses on primary microplastics, secondary microplastics are rarely studied. Microplastics have been found in many foods, but little is known about their bioaccumulation and health risks. Future studies should focus on the health risks of microplastics in wastewater, assessing their hazards and exploring ways to reduce them. Long-term studies are needed to determine the impact of microplastics in wastewater on aquatic ecosystems and the environment, which will help to develop effective management and mitigation methods. There are few comprehensive studies on the management and elimination of microplastics in wastewater. Effective remediation requires further research.

Global collaboration in research is possible, especially among leading research nations like China, the United States, and the United Kingdom, as well as emerging academic regions. A significant number of countries (129 out of 149) have limited studies on microplastics, pointing out the global disparity in research and its untapped potential. It is crucial to bolster research capabilities in these overlooked nations. Top institutions from these regions can play a role in establishing new research centers. Expert researchers like J. Wang offer valuable guidance in the field. Their methods could benefit newcomers. Frequently cited studies shed light on research priorities and areas of interest. There's an urgent need to delve into the origins of microplastics, their effects in unexplored regions, and ways to counteract them. For better precision in results, future studies should refine the methodologies proposed by Hidalgo-Ruz et al. (2012). Even though countries like the Netherlands and the United Kingdom aren't major contributors in terms of volume, their high citation indices underline the significance of research quality. Comprehensive studies on microplastics, spanning environmental, health, and engineering domains, are essential. Publishing this research in esteemed journals can help shape policies and best practices.

ACKNOWLEDGMENTS

The authors thank Asso. Prof. Dr. I. Panigrahi and Dy. Librarian Dr. B.B.K. Patra from KIIT University School of Mechanical Engineering, Bhubaneswar, Odisha for their assistance with data collection.

AUTHORS' CONTRIBUTIONS

S.K. Sahu and K. Murugesan came up with the idea for the article; B.S. Panda, R.K. Sethi, and M. Yadav conducted the literature survey; S.S. Pati, B.S. Panda, and A. Sahu conducted the data collection and analysis; and K. Murugesan and S.K. Sahu have critically revised the work. All authors read and approved the final manuscript.

REFERENCES

Akinpelu, E. A., & Nchu, F. (2022). A bibliometric analysis of research trends in biodegradation of plastics. *Polymers*, *14*(13), 2642.

Alimi, O. S., Farner Budarz, J., Hernandez, L. M., & Tufenkji, N. (2018). Microplastics and nanoplastics in aquatic environments: Aggregation, deposition, and enhanced contaminant transport. *Environmental Science & Technology*, *52*(4), 1704–1724.

Auta, H. S., Emenike, C. U., & Fauziah, S. H. (2017). Distribution and importance of microplastics in the marine environment: A review of the sources, fate, effects, and potential solutions. *Environment International*, *102*, 165–176.

Bhangare, R. C., Tiwari, M., Ajmal, P. Y., Rathod, T. D., & Sahu, S. K. (2023). Exudation of microplastics from commonly used face masks in COVID-19 pandemic. *Environmental Science and Pollution Research, 30*(12), 35258–35268.

Borrelle, S. B., Ringma, J., Law, K. L., Monnahan, C. C., Lebreton, L., McGivern, A., Murphy, E., Jambeck, J., Leonard, G. H., Hilleary, M. A. & Rochman, C. M. (2020). Predicted growth in plastic waste exceeds efforts to mitigate plastic pollution. *Science, 369*(6510), 1515–1518.

Carr, S. A., Liu, J., & Tesoro, A. G. (2016). Transport and fate of microplastic particles in wastewater treatment plants. *Water Research, 91*, 174–182.

Chae, Y., & An, Y. J. (2017). Effects of micro- and nanoplastics on aquatic ecosystems: Current research trends and perspectives. *Marine Pollution Bulletin, 124*(2), 624–632.

Chang, X., Fang, Y., Wang, Y., Wang, F., Shang, L., & Zhong, R. (2022). Microplastic pollution in soils, plants, and animals: A review of distributions, effects and potential mechanisms. *Science of the Total Environment, 850*, 157857.

Chang, X., Xue, Y., Li, J., Zou, L., & Tang, M. (2020). Potential health impact of environmental micro-and nanoplastics pollution. *Journal of Applied Toxicology, 40*(1), 4–15.

Cox, K. D., Covernton, G. A., Davies, H. L., Dower, J. F., Juanes, F., & Dudas, S. E. (2019). Human consumption of microplastics. *Environmental Science & Technology, 53*(12), 7068–7074.

Dris, R., Gasperi, J., Saad, M., Mirande, C., & Tassin, B. (2016). Synthetic fibers in atmospheric fallout: A source of microplastics in the environment? *Marine Pollution Bulletin, 104*(1–2), 290–293.

Fendall, L. S., & Sewell, M. A. (2009). Contributing to marine pollution by washing your face: Microplastics in facial cleansers. *Marine Pollution Bulletin, 58*(8), 1225–1228.

Free, C. M., Jensen, O. P., Mason, S. A., Eriksen, M., Williamson, N. J., & Boldgiv, B. (2014). High-levels of microplastic pollution in a large, remote, mountain lake. *Marine Pollution Bulletin, 85*(1), 156–163.

Gall, S. C., & Thompson, R. C. (2015). The impact of debris on marine life. *Marine Pollution Bulletin, 92*(1–2), 170–179.

Gao, Y., Ge, L., Shi, S., Sun, Y., Liu, M., Wang, B., Shang, Y., Wu, J. & Tian, J. (2019). Global trends and future prospects of e-waste research: A bibliometric analysis. *Environmental Science and Pollution Research, 26*, 17809–17820.

Geyer, R., Jambeck, J. R., & Law, K. L. (2017). Production, use, and fate of all plastics ever made. *Science Advances, 3*(7), e1700782.

Global plastic production 1950-2021. Statista Research Department, 2023. https://www.statista.com/statistics/282732/global-production-of-plastics-since-1950/#:~:text=Global%20plastic%20production%201950%2D2021&text=Global%20plastics%20production%20was%20estimated,production%20has%20soared%20since%201950s.

Hartmann, N. B., Huffer, T., Thompson, R. C., Hassellov, M., Verschoor, A., Daugaard, A. E., Rist, S., Karlsson, T., Brennholt, N., Cole, M. & Wagner, M. (2019). Are we speaking the same language? Recommendations for a definition and categorization framework for plastic debris. *Environmental Science & Technology*, 53(3), 1039–1047.

Hernandez, E., Nowack, B., & Mitrano, D. M. (2017). Polyester textiles as a source of microplastics from households: A mechanistic study to understand microfiber release during washing. *Environmental Science & Technology, 51*(12), 7036–7046.

Hidalgo-Ruz, V., Gutow, L., Thompson, R. C., & Thiel, M. (2012). Microplastics in the marine environment: A review of the methods used for identification and quantification. *Environmental Science & Technology, 46*(6), 3060–3075.

Horton, A. A., Walton, A., Spurgeon, D. J., Lahive, E., & Svendsen, C. (2017). Microplastics in freshwater and terrestrial environments: Evaluating the current understanding to identify the knowledge gaps and future research priorities. *Science of the Total Environment, 586*, 127–141.

Kumar, R., Sharma, P., Manna, C., & Jain, M. (2021). Abundance, interaction, ingestion, ecological concerns, and mitigation policies of microplastic pollution in riverine ecosystem: A review. *Science of the Total Environment, 782*, 146695.

Liu, W. H., Zheng, J. W., Wang, Z. R., Li, R., & Wu, T. H. (2021). A bibliometric review of ecological research on the Qinghai–Tibet Plateau, 1990-2019. *Ecological Informatics, 64*, 101337.

Ma, B., Xue, W., Hu, C., Liu, H., Qu, J., & Li, L. (2019). Characteristics of microplastic removal via coagulation and ultrafiltration during drinking water treatment. *Chemical Engineering Journal, 359*, 159–167.

Mintenig, S. M., Int-Veen, I., Loder, M. G., Primpke, S., & Gerdts, G. (2017). Identification of microplastic in effluents of waste water treatment plants using focal plane array-based micro-Fourier-transform infrared imaging. *Water Research, 108*, 365–372.

Murphy, F., Ewins, C., Carbonnier, F., & Quinn, B. (2016). Wastewater treatment works (WwTW) as a source of microplastics in the aquatic environment. *Environmental Science & Technology, 50*(11), 5800–5808.

Oginni, O. (2022). COVID-19 disposable face masks: A precursor for synthesis of valuable bioproducts. *Environmental Science and Pollution Research, 29*(57), 85574–85576.

Padilla, F. M., Gallardo, M., & Manzano-Agugliaro, F. (2018). Global trends in nitrate leaching research in the 1960–2017 period. *Science of the Total Environment, 643*, 400–413.

Papadimitriu, M., & Allinson, G. (2022). Microplastics in the Mediterranean marine environment: A combined bibliometric and systematic analysis to identify current trends and challenges. *Microplastics and Nanoplastics, 2*(1), 1–25.

Renzi, M., Grazioli, E., & Blašković, A. (2019). Effects of different microplastic types and surfactant-microplastic mixtures under fasting and feeding conditions: A case study on Daphnia magna. *Bulletin of Environmental Contamination and Toxicology, 103*(3), 367–373.

Schwabl, P., Koppel, S., Königshofer, P., Bucsics, T., Trauner, M., Reiberger, T., & Liebmann, B. (2019). Detection of various microplastics in human stool: A prospective case series. *Annals of Internal Medicine, 171*(7), 453–457.

Sher, F., Hanif, K., Rafey, A., Khalid, U., Zafar, A., Ameen, M., & Lima, E. C. (2021). Removal of micropollutants from municipal wastewater using different types of activated carbons. *Journal of Environmental Management, 278*, 111302.

Van Cauwenberghe, L., & Janssen, C. R. (2014). Microplastics in bivalves cultured for human consumption. *Environmental Pollution, 193*, 65–70.

Van Sebille, E., Wilcox, C., Lebreton, L., Maximenko, N., Hardesty, B. D., Van Franeker, J. A., Eriksen, M., Siegel, D., Galgani, F. and Law, K. L. & Law, K. L. (2015). A global inventory of small floating plastic debris. *Environmental Research Letters, 10*(12), 124006.

Vianello, A., Jensen, R. L., Liu, L., & Vollertsen, J. (2019). Simulating human exposure to indoor airborne microplastics using a Breathing Thermal Manikin. *Scientific Reports, 9*(1), 8670.

Wagner, C. S., Park, H. W., & Leydesdorff, L. (2015). The continuing growth of global cooperation networks in research: A conundrum for national governments. *PloS one, 10*(7), e0131816.

Walker, T. R., & Fequet, L. (2023). Current trends of unsustainable plastic production and micro (nano) plastic pollution. *Trends in Analytical Chemistry, 160*, 116984.

Xu, E. G., & Ren, Z. J. (2021). Preventing masks from becoming the next plastic problem. *Frontiers of Environmental Science & Engineering, 15*(6), 125.

Zhang, S., Wang, J., Liu, X., Qu, F., Wang, X., Wang, X., Wang, X., Li, Y. & Sun, Y. (2019). Microplastics in the environment: A review of analytical methods, distribution, and biological effects. *Trends in Analytical Chemistry, 111*, 62–72.

Zhou, C., Bi, R., Su, C., Liu, W., & Wang, T. (2022). The emerging issue of microplastics in marine environment: A bibliometric analysis from 2004 to 2020. *Marine Pollution Bulletin, 179*, 113712.

2 Characterization Techniques for Quantifying Environmental Microplastics

Devansh Sharma and Nallin Sharma

2.1 INTRODUCTION

According to the United Nations Environment Program (UNEP), the amount of plastic waste has increased 200-fold since 1950 (Cole et al., 2011). With increasing plastic waste and improper handling, this plastic waste often ends up in water bodies or landfills where it can remain for a longer time. This plastic waste breaks down into microplastics under the action of different environmental factors such as UV radiation and thermal cracking, and thus considered secondary microplastics (da Costa et al., 2016). The size of microplastic may range from <1 μm to 5 mm and are often referred to as "Mermaid Tears. Microplastics can be sub classified as primary or secondary microplastics" (Cole et al., 2011). The different sources of primary and secondary microplastics are shown in Table 2.1. Primary microplastics are referred to the small plastic particles that are available in daily products such as toothpaste and shampoo, which leach out to the water bodies. Secondary microplastics are the result of continuous breaking of plastic waste into smaller particles by the action of mechanical churning between the sea and sand particles or due to the degradation of plastic waste. These microplastic particles further degrade due to the action of biotic and abiotic factors. The presence of microplastic is also observed in deep sea, land, and air (Barrett et al., 2020; Shaw & Day, 1994).

Overall, PL spectroscopy provides efficient methods for the detection of microplastics in different environments. This characterization tool enables us to identify a wide range of microplastics in the ambient environment. Moreover, this method requires an inexpensive characterization tool as PL uses a visible light source. PL, coupled with other characterization tools such as FTIR, also expands its horizon in qualitative and quantitative analysis of microplastics.

Recently, researchers have observed the presence of microplastics in our diaspora at different levels and have identified it as a potential threat to the environment (da Costa et al., 2016). Microplastics are present in different shapes, sizes,

DOI: 10.1201/9781003438793-2

TABLE 2.1
Sources of Different Microplastic in the Environment

Sources of Primary and Secondary Microplastics	
Primary microplastics	Personal care products such as exfoliants
	Specific medical products, including dental tooth polish
	Industrial Abrasive
	Drilling fluids
	Raw materials for plastic production, process subproducts
Secondary microplastics	General littering, dumping of plastic waste
	Abrasion in landfill sites and recycling facilities
	Carelessly handled plastic fishing gear
	Ship-generated litter or disposed of after recreational activities
	Plastic material present in organic waste
	Polymer used in compositing additives
	Fibers released from hygiene product
	Fibers released from synthetic textile

Source: As reviewed by K. Duis, A. Coors, *Environ. Sci. Eur.* 28(1), 2016.

and structures (Phuong et al., 2016). This has made the identification and analysis of microplastics a challenging job. Moreover, these microplastics are found in combination with different additives and chemicals, thus making it a challenge to be identified and analyzed (Ivleva, 2021).

2.2 CHARACTERIZATION OF MICROPLASTICS

The characterization of MPs involves their physical and chemical compositions. Physical characterization is concerned with the identification of physical factors such as shape, size, morphology, and color, whereas chemical analysis is related to the identification of chemical bonds and functional groups. Various characterizing techniques for microplastics are discussed below.

2.2.1 FTIR

It is known that every polymer represents the specific FTIR spectrum. The polymers are characterized by spectral peaks in the fingerprint region. This region possesses peaks associated with the functional groups present. In general, unknown components in a sample can be identified by interpreting FTIR spectra and assigning the vibrational bands to transitions of functional groups (Reddy et al., 2006). However, this technique is time-consuming. In another technique, a more practical approach is taken where the obtained spectrum is compared with a reference database (consisting of characteristic spectra of different plastics). The MP spectra are unidentical to its parent polymers as a result of its manufacturing processes or its interactions with the environment (Fotopoulou & Karapanagioti, 2012). However, some characteristic peaks match with the parent polymer, thus helping in characterizing microplastic samples.

In the general procedure, the collected IR spectra are processed through smoothing of graph using *Savitzky-Golay* algorithm, and subtraction of baseline. Thus, obtained spectra is compared with the reference for the identification of microplastics.

2.2.1.1 Application

FTIR is used to analyze microplastic samples in sediments (e.g., Soil), floating (e.g., Water bodies), and ingested (e.g., Fish) (Andrady, 2011; Cole et al., 2011; Filella, 2015; Hidalgo-Ruz et al., 2012). FTIR is commonly used to identify and differentiate microplastics such as polyethylene, polypropylene, and polyamide. FTIR is a known and reliable method for characterizing microplastic in various situations. Claessens et al. were able to distinguish several microplastics by FTIR and ATR techniques in different environments such as sediments of coastal areas including sea beds and beaches (2011). They were able to analyze microplastic samples with different shapes and sizes, for example, fibers, granules, plastic films, and spherules.

Harrison et al. (2012) used ATR and reflectance method to identify polyethylene microplastic over a membrane filter. They established the morphology of the microplastic results in scattering interferences and recommended the use of ATR technique. They further did chemical imaging of the sample by using an FPA detector. Tagg et al. tried to analyze different plastics together by combining FT-MIR microscopy and FPA technique to analyze six different plastics in a sample (polyethylene (PE), polypropylene (PP), polyamide 6 (PA6), polystyrene (PS) and polyvinyl chloride (PVC)) by using FT-MIR microscopy and FPA detector (2015). In their research, they also treated wastewater with H_2O_2 to remove microorganisms and then analyzed for microplastics in wastewater. This technique resulted in improved characterization of most common microplastics such as PE (0.4 particles/L) and PVC (0.2 particles/L).

The third prominent way to analyze microplastics is based on ingestion. Frias et al. studied zooplankton in Portuguese coastal water for microplastics (2014). They found microplastics in about 61% of collected zooplankton. They contain polyethylene, PP, and polyacrylate. This raises the concern about microplastics entering the food chain. In their studies, Lusher et al. (2013) and Foekema et al. (2013) analyzed the microplastics indigested by fish. Lusher identified polyamide in six microplastic samples in 36% of fish, whereas Foekma detected polyethylene and polypropylene. All these become carriers for microplastics to the deep sea.

2.2.2 Raman Spectroscopy

Raman spectroscopy is another method used to identify the polymer composition of microplastics. It involves analyzing the scattering of laser light by the particles, which provides characteristic molecular vibrational spectra for identification. Raman spectroscopy is a very potent tool to analyze microplastic. It resolves the issue of diversified environmental conditions; moreover, it can be used to analyze a broad spectrum of particle size (1 μm) and provide an efficient method to estimate the chemical composition. It also enables to trace microplastics at the subcellular level present in living microorganisms.

Raman spectroscopy has some advantages over other tests as it provides (i) broader spectral coverage, (ii) higher resolution, and (iii) higher moisture tolerance. Raman analysis is preferred to analyze samples with high moisture content or with water.

As water molecules are highly polar, therefore, the polarizability of water changes very low during vibration, therefore the Raman spectrum for the water is very weak. Thus, it is preferred for wet samples. Moreover, the spatial resolution of Raman microscopy is typically ca. 1 μm, which enables us to analyze microplastic sample >1 μm in size in comparison to IR microscopy, which can allow particles with 10–20 μm (Imhof et al., 2012; Lenz et al., 2015). Since, in Raman spectroscopy, lasers are used, which have the potential to probe a smaller region and can be performed under different ambient conditions, even for living organisms (Smith et al., 2016).

Raman imaging is known to identify different phases in a sample. In many cases, Raman imaging helped in the analysis of different polymers in a mixture by Raman mapping. Raman imaging supported with other characterizations is often used to analyze phases at nanoscales such as Raman-AFM (Schmid et al., 2013). In another approach, confocal Raman analysis is used to generate 3D images of microplastics at the subcellular level inside living organisms (Cole et al., 2013; Watts et al., 2014).

The Raman analysis uses laser, which can degrade the sample over a long exposure. Another problem associated with Raman analysis is the fluorescence (intrinsic or caused by impurities) overlying Raman spectra. This brings another issue associated with analysis of microplastics is the analysis of colored plastics. Raman analysis has been proven to be an efficient method for detecting and analyzing microplastics in different environment, such as in sediments (Fischer et al., 2015; Imhof et al., 2012, 2016; Van Cauwenberghe et al., 2013, 2015, Zhang et al., 2016) and 29% water samples both seawater (Enders et al., 2015; Löder & Gerdts, 2015; Zettler et al., 2013) and freshwater (Imhof et al., 2013; Zhao et al., 2015).

Another advancement in Raman analysis is the use of two coherent sources for the irradiation of samples (CARS). It is much preferable over Raman analysis as it provides easy sample preparation. The problem of fluorescence can also be overcome by using CARS spectroscopy. Likewise, Cole et al. (2013) were able to detect small microplastic beads of PS (0.4–30 um) inside the copepods using CARS. It uses the characteristic band of PS at 2,845/cm, signifying C-H stretching, and 3,050/cm, from C-H aromatic moieties. The contrast difference appears where strong anti-Stokes signals are generated and are marked in false color, thus helping in the mapping of microplastics inside the organism. In a similar way, Watts and coworkers studied the microplastic uptake in shore crabs (Watts et al., 2014).

In recent years, Raman spectroscopy has been proven to be an excellent tool to identify microplastic pollutants. However, there is a need for a standardized protocol for sample collection and acquisition of spectra. There is a need to extend the range of microplastic detection under 1 mm which consists of 100 particles of 1 μm. In the future, we will be witnessing the rise of these microparticles. We must reinvent the method of separation in accordance with the decreasing particle size. The MPSS apparatus designed by Imhof et al. (2012) is a crucial step in this direction. In recent studies it has been observed that visual inspection of sampling gives unreliable results with false positive and false negative. There is a need to identify and confirm the identity of microplastic through available methods like mass spectroscopy, vibrational spectroscopy, or a combination. Another crucial step is the scale-up process to handle large particles. The current technique CARS has an edge over other methods as it allows imaging of all particles of similar type all at once. There are some limitations associated with the Raman analysis, as it uses laser, which can degrade the sample over long exposure. Another problem associated with Raman analysis is the fluorescence (intrinsic or caused by impurities) overlying Raman spectra. This brings another issue associated with the analysis of microplastics is the analysis of colored plastics.

2.2.3 Photoluminescence Spectroscopy (PL)

Photoluminescence is a material property that emits light when irradiated with a photon. This property is shown by some molecules which emit photons upon irradiation. Microplastics do not show significant photoluminescence; therefore, they are often detected using a staining agent. Different staining agents have their respective PL absorption spectra. These staining agents are small molecules often known as dyes. These dyes are attached to microplastics; thus, PL spectra can easily distinguish different microplastics. This method provides a fast method for the detection of microplastics. There are many such molecules like Nile Red, etc. (Maes et al., 2017). These molecules are often used as labeling agents; therefore, they are selectively attached to the material to be detected under PL spectroscopy. This approach has been utilized in the detection of microplastics in different environments (Konde et al., 2020).

The method of time-resolved PL spectroscopy (TR-PL) has also been explored in the field of microplastic detection (Gies et al., 2020). With the use of TR-PL, they were able to identify different types of plastics and differentiate between different impurities such as sand particles and microplastics. In another approach, Sharma et al. (2021) used engineered nanoparticles for the detection of microplastics, as represented in Figure 2.1. In their approach, they were able to detect PAN microplastics

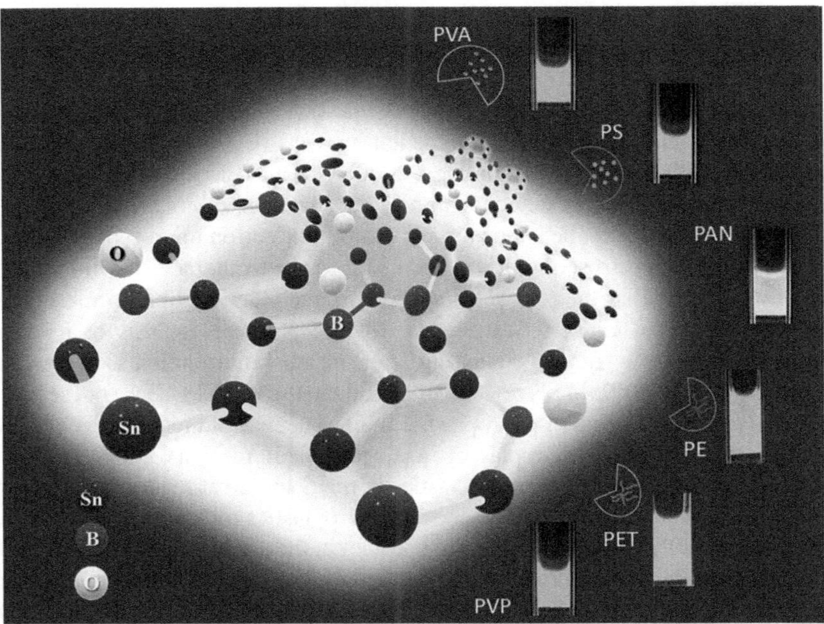

FIGURE 2.1 Synthesis of Stanene oxyborohydride (SnOB) nanosheets (Reproduced from Sharma et. al., 2021).

with a smaller size <5 μm, a limit at par with the current technological benchmark. They used stanene oxyboride nanosheets to detect PAN microplastics. They studied the change in PL properties of these nanosheets in the presence of different microplastics, which helps in probing the microplastics. Also, this methodology enables us to detect different microplastics. Such a method provides label free detection of microplastics without the use of dyes.

2.2.4 Microscopy

Microscopy is a tool to produce an image of a 3D object and then magnify the image. Thus, presenting the magnified image of the smaller objects that are unresolvable from the human eye. The aim of microscopy is to provide better-magnified images for smaller objects. With advancements in microscopy, we can explore and study fundamental properties at subatomic level. For years, optical-light microscopy (OM) was the easily available magnifying imaging system. In this method, objects are illuminated with light sources such as halogen lamps, lasers, LEDs, or focused natural light. With the help of optics (lenses, glasses, etc.), the light is focused on the sample. The focused image is formed on the objective lens. This image is further magnified by the ocular lens, giving a final magnified image of the specimen. Theoretically, the final magnified image is the product of the power of the ocular and the objective lenses (Croft, 2006; Egerton, 2005; Haynes, 2013; Török & Kao, 2007).

There are several limitations to using OM, such as deterioration of resolution with increasing magnifications. There has been extensive research in the field of microscopy to overcome the restriction of the OM. Electron microscopy (EM) is one such method that uses a high-energy electron beam as a source, which theoretically gives a wavelength of irradiation between 0.001 and 0.01 nm. Thus, EM provides a theoretical resolution calculated as 0.02 nm, for a 100 kV electron beam. In EM, we use magnetic lenses that focus the electron beam to subatomic size, thus enabling it to reach very high resolution and also providing elemental analysis (Egerton, 2005; Goodhew & Humphreys, 2000; Reimer & Technology, 2000). OM and EM are often used complementary to each other in the detection and characterization of microplastics.

This chapter discusses the application of scanning electron microscopy (SEM) and energy dispersion spectroscopy (EDS) and its relevance to characterize microplastics. Finally, we will discuss the role of microscopy in the identification of microplastics. SEM-EDS is a vital characterization tool for the physical analysis of microplastic samples, especially those originating from marine environments. For example, Hidalgo-Ruz et al. (2012) in his review discussed the use of SEM-EDS for the identification and characterization of microplastics in marine environments. SEM has been a prominent tool to identify microplastics in aqueous medium. Van Cauwenberghe et al. (2013) showed the presence of microplastics in the deep sea at the top sediment layer over the sea bed, signifying the presence of microplastics all over the sea. Similarly, Erikssen et al. (2013) found microplastic particles at the surface of the lake. Weinstein et al. (2016) established the role of UV radiation in the degradation of plastic debris into microplastics. Researchers have also used SEM-EDS to analyze and identify inorganic compounds in microplastic samples present in additives added to parent plastics (Fries et al., 2013; Rios Mendoza & Jones, 2015).

SEM technique has also been used by researchers to establish the interaction between biota and microplastics. Like Carson et al. (2013) proved the bacterial growth in the presence of the rough surface such as microplastic. Similarly, Zettler et al. (2013) investigated the growth of microbial communities over plastic debris and concluded that latter are the ecological habitat for the diverse microbial communities such as heterotrophs, autotrophs, predators, and symbionts, which have been referred to as *Plastisphere*. Through different studies done using SEM, researchers have confirmed the interaction between organisms and microplastics. According to their research, microplastics provide a habitat for bacterial growth.

Use of conventional Transmission Electron Microscopy (TEM) to analyze microplastic samples is practically impossible as it operates at accelerating voltages of 80, 200, or 300 kV. At such high voltage, it is impractical to observe microplastics without destroying them. A recent study used low voltage TEM (5 kV)to analyze nano-sized plastics: Gigault et al. (2016) reported the nanoplastics to exceed the size of 100 nm due to degradation of microplastics under UV. Similarly, Booth et al. (2013, 2016) studied both polymethylmethacrylated (PMMA) based nanoplastic synthesis and its influence on different aquatic organisms. In many reviews, researchers have highlighted the presence of nanoplastics in the biosphere, their sources, fates, and effects (Andrady, 2011; da Costa et al., 2016).

2.2.5 Pyrolysis Gas Chromatography/Mass Spectrometry (Py-GC/MS)

Information about the chemical composition of microplastics is crucial. There are different techniques that have been explored to perform the chemical analysis of the polymer sample such as Fourier Transform Infrared (FTIR) spectroscopy, Raman spectroscopy, and analytical pyrolysis coupled with gas chromatography/mass spectrometry (Py-GC/MS) (Hidalgo-Ruz et al., 2012; Löder & Gerdts, 2015; Rocha-Santos & Duarte, 2015). In this method, we compare the obtained sample spectrum with the reference spectra of known polymers, thus providing information about the available polymers. This method allows us to investigate different polymers. In this method, the polymer sample is heated at the elevated temperature of 500°C–1,400°C under a helium gas environment. The decomposed products are collected using a fused silica capillary column, detected, and analyzed by mass spectrometer; further, the received spectra are interpreted using mass spectral libraries (e.g., NIST or Wiley). The advantage of using the pyrolysis method is that there is no requirement for the pretreatment of the samples. Also, this methodology is known to give reliable results irrespective of the shape of the sample. There are two different methods used to analyze pyrolyzed samples: (i) Py-GC/MS and (ii) TED (thermo-extraction desorption) GC/MS. There are different pyrolysis methods that are applied during Py-GC/MS. These modes and their application in the determination of microplastics are extensively reviewed by Ivleva (2021).

2.2.6 Fluid Imaging Flow Cytometry

However, the microplastic analysis can be done qualitatively and quantitatively using FTIR and Raman. However, both the samples required special apparatus and sample preparation, which is time-consuming and not commercially viable. Therefore, to address the challenge of rapid analysis of microplastics in wastewater, Hyeon et al. (2023) explored the use of flow cytometry and fluid imagining. Researchers have also used imaging software(Yang et al., 2019) or gravimetric-based estimations(Napper & Thompson, 2016; Pirc et al., 2016) to approximate the quantity of microfiber present in the sample. Researchers are looking into the possibility of adopting automated or semi-automated data collection technology to overcome the issues.

In this pursuit, Hyeon et al. (2023) used fluid imaging flow cytometry and studied various properties of microplastic fibers. They studied the flow behavior of microplastic fibers in silicon carbide and alumina ceramic membranes after the treatment of laundry wastewater. They were successfully able to establish that the morphology of the fibers affects the rentability of the membrane. According to them, long and curly fibers are retained in the ceramic filter. Also, they found that the number of microfibers increases with increasing laundry. Smaller particles, like spherical-shaped particles or relatively compressed microplastic fiber, were found to permeate through the membrane. Thus, this method proves to be an efficient and meaningful method to analyze and image the microplastic in the flowing water resources. These results are important as no other technique can provide this analysis. The only limitation to its application is that microplastic pollutant with particle size <50 um passes through the membrane under pressure, thus making it inefficient for smaller particle sizes.

2.2.7 Laser-Induced Breakdown Spectroscopy

Laser-induced breakdown spectroscopy (LIBS) is an analytical tool for analyzing microplastics and helps in fast and in situ measurement of the microplastic, delivering accurate findings for identifying the kind of polymer and other additives contained in the samples. This is a proven methodology in different strata (Dockery & Goode, 2003), food analysis (Agrawal et al., 2011), underwater sample analysis (Guo et al., 2017), and even in space (De Giacomo et al., 2007). The information provided by this analysis is based on the chemical constituents and molecular structure of the sample. In this technique, the laser is fired on the sample, creating plasma. After the plasma cools down, elements are characterized by the characteristic radiation emitted, which is further detected by the spectrometer. There are several advantages of using LIBS, such as there is no sample preparation required, thus allowing in situ measurements; it can also analyze samples independent of their physical state; this method is also insensitive to the surface morphology, and thus, can help in the detection of surface contaminants.

LIBS has been recently exploring the field of microplastic detection. Chen et al. did compositional studies in single non-plastic microparticles (Chen, Huang, et al., 2020), and they also determine the concentration of heavy metal in microplastics from river samples (Chen, Wang, et al., 2020). Sommer et al. (2021) used LIBS to distinguish microplastics from natural products and to identify base polymer types, as represented in Figure 2.2. This methodology combines results after line intensities from different elements and multivariate analysis. He also proves that LIBS also predicts microplastic particles even if the polymer matrix is contaminated or oxidized. In conclusion, this technique offers several advantages over other characterization

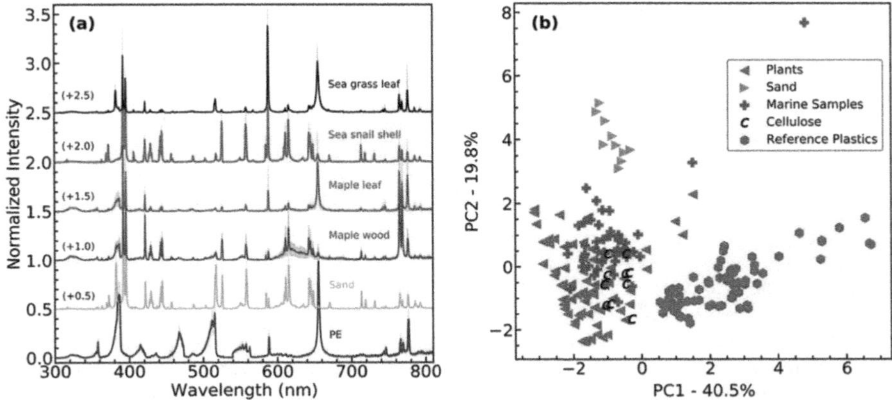

FIGURE 2.2 Identification of different microplastic samples using LIBS. This also helps in the determination of other impurities and different types of microplastics (Sommer et al., 2021). Reproduced with permission from Sommer et al., 2021.

tools such as instant identification (1 second per scan), mobility, and depth profile identification. It helps in identifying toxic pollutants with minimal destruction to sample with the application of less number of laser shots (up to single shot LIBS) and with the reduced laser power (thus the depleted amount ranges from tens to hundreds of nanograms) (Gaudiuso et al., 2010). Thus proving it to be an efficient method for the detection of microplastics.

2.2.8 Light Scattering and Chemometric

Light scattering is a known technique used for particle analysis (Dick et al., 2007). Light scattering phenomena can be subdivided into static and dynamic. In static scattering, the intensity of light is a function of the angle of incidence, whereas in dynamic scattering, it is the function of time (Hodgson, 2001).

In recent research, transparent and translucent MPs were detected in freshwater by using a light scattering experiment (Asamoah et al., 2019, 2020). However, determining the size of nano plastic is not reliable with this method. Therefore, different modifications have been suggested to traditional light scattering experiments. Caputo et al. (2021) investigated nine different methods to analyze the size and mass of polystyrene (PS) particles and concluded that the light scattering method is inefficient in resolving different particles in polydisperse samples. In another approach, Huang et al. (2018) employed an integrated microfluidic chip with an on-chip multiangle laser scattering module and a preconcentration step to detect waterborne parasites. With the proposed method, they were able to distinguish 2 and 10 μm PS particles. Choobbari et al. (2022) used multiangle static light scattering analyzer, called a goniophotometer, along with principal component analysis (PCA) to extract the Zernikr moment features of scattering pattern and linear discriminant analysis (LDA) algorithm for the classification of the patterns as represented in Figure 2.3.

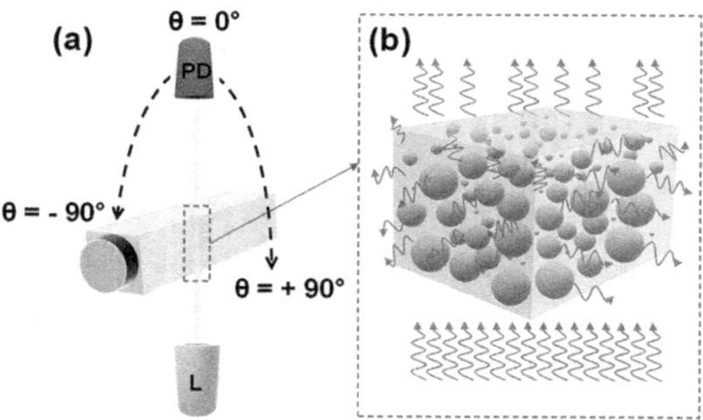

FIGURE 2.3 Schematic representation of light scattering and chemometric technique for identification of microplastics (a) "PD" and "L" refers to photodiode and light source, respectively. (b) Demonstrates the scattering of light by MPs. Reproduced with permission from Choobari et al.

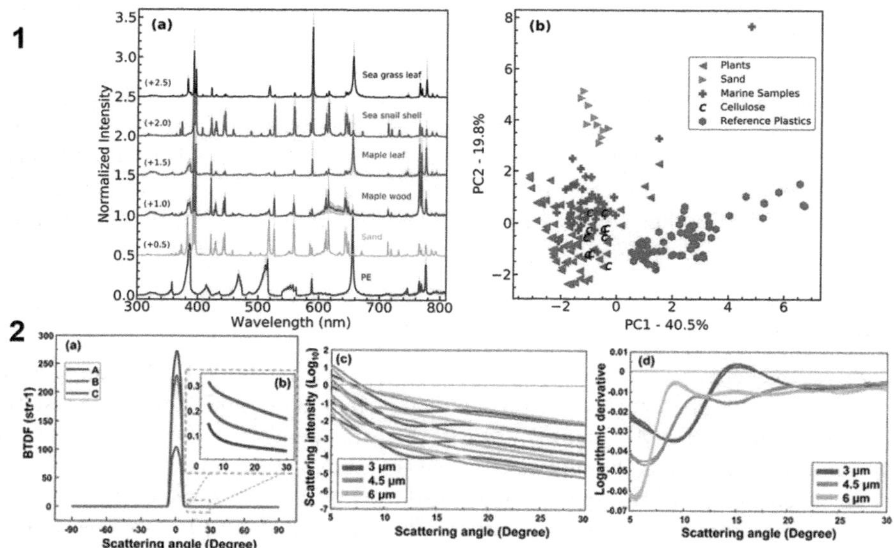

FIGURE 2.4 Identification of microplastics using light scattering and chemometric.

They were able to determine MP size and concentration in a batch of both mono- and polydisperse samples because to this technique. The author estimates that the PCA-LDA model can be applied to a polydisperse mixture, and it became crucial for the unknown size of microplastic in a mixture, as represented in Figure 2.4.

2.3 LIMITATION AND CHALLENGES

The characterization of microplastics poses several limitations and challenges, which can impact the accuracy and reliability of the results. Some of the key limitations and challenges associated with microplastic characterization are:

a. Size Range: Microplastics are defined as particles smaller than 5 mm, encompassing a wide range of sizes from nanometers to millimeters. Analyzing microplastics at both ends of the size spectrum can be challenging, as different techniques may be required to capture and measure particles within this range. Thus, a versatile approach with the capacity of broad size analysis is desirable.
b. Sample Preparation: Microplastics are often present in complex matrices, such as sediments, water, or biological tissues, which can make their extraction and isolation laborious and time-consuming. Sample preparation techniques may introduce biases or loss of microplastics, affecting the accuracy of the characterization.
c. Identification of Polymer Types: Identifying the specific polymer types of microplastics can be challenging, especially when they are weathered or fragmented. Over time, microplastics can undergo physical and chemical changes, making it more difficult to determine their original polymer composition.
d. Fragmentation and Degradation: Microplastics can undergo fragmentation and degradation processes in the environment, leading to changes in their size, shape, and chemical composition. Characterizing degraded microplastics or distinguishing between primary and secondary microplastics can be challenging, as they may exhibit different properties and behaviors.
e. Contamination: Microplastics are susceptible to contamination during sampling, handling, and laboratory analysis. Contamination sources can include airborne particles, glass-based materials used in lab equipment, or cross-contamination from other samples. Controlling and minimizing contamination is crucial to obtain accurate results.
f. Standardization and Reference Materials: The absence of standardized methods and reference materials for microplastic analysis can pose challenges in comparing and interpreting data between studies. Establishing standard protocols and reference materials would enhance the accuracy and reproducibility of microplastic characterization.
g. Lack of Quantification Methods: Accurately quantifying microplastics in environmental samples is challenging due to their low concentrations and the presence of interfering particles. Developing robust and sensitive quantification methods that account for sample matrix effects and interferences is essential for accurate assessment of microplastic abundance.
h. Data Interpretation and Reporting: The interpretation and reporting of microplastic data require careful consideration of uncertainties, limitations, and potential biases associated with the chosen analytical techniques. Transparent reporting of methods, quality control measures, and uncertainties is necessary to ensure data reliability and comparability.

Addressing these limitations and challenges requires ongoing research, method development, collaboration among scientists, and continuous improvement in analytical techniques. It is essential to adopt standardized protocols, improve data comparability, and develop innovative approaches to overcome these challenges and advance our understanding of microplastics and their environmental impact.

2.4 FUTURE PROSPECTIVE

The prospects of microplastic detection are focused on advancing and developing more sensitive, efficient, and comprehensive methods to detect and characterize microplastics. Here are some potential areas of development:

a. Improved Analytical Techniques: Research and development efforts are ongoing to enhance existing analytical techniques such as spectroscopy (FTIR and Raman), microscopy (SEM and TEM), and chromatography (Py-GC/MS) to improve their sensitivity, speed, and accuracy. This includes advancements in miniaturized and portable instrumentation for on-site analysis.
b. Nanotechnology and Sensor Development: There is growing interest in developing nanotechnology-based sensors and devices specifically designed for microplastic detection. These sensors may utilize various principles, such as surface-enhanced Raman spectroscopy (SERS) or nanomaterial-based sensors, to detect and quantify microplastics at low concentrations.
c. Molecular Techniques: Molecular techniques, including DNA-based methods, are being explored to identify and characterize microplastics. These techniques involve the extraction and amplification of genetic material associated with microplastics, allowing for species-specific identification and potentially providing insights into their sources and impacts.
d. Machine Learning (ML) and Artificial Intelligence (AI): The application of ML and (AI) algorithms can enhance the efficiency and accuracy of microplastic detection and classification. These techniques can assist in automated image analysis, spectral analysis, and data processing, enabling rapid and high-throughput analysis of large datasets.
e. Microplastic Surface Analysis: Understanding the surface properties of microplastics is crucial in assessing their environmental fate and interactions. Research is ongoing to develop techniques that can provide detailed surface characterization, including surface charge, functional groups, and adsorbed contaminants.
f. Standardization and Harmonization: There is a need for standardized protocols and methods for microplastic detection and analysis to ensure comparability of data across different studies. Efforts are being made to establish standardized sampling, sample processing, and analysis procedures to facilitate more robust and consistent microplastic research.
g. Real-time Monitoring: Advancements in sensor technology may enable real-time monitoring of microplastics in environmental systems, such as water bodies or wastewater treatment plants. Real-time monitoring can provide valuable insights into the temporal and spatial distribution of microplastics and their sources.
h. Integrated Multitechnique Approaches: The future of microplastic detection lies in the integration of multiple techniques, combining physical, chemical, and biological approaches. By employing complementary methods, researchers can obtain a more comprehensive understanding of microplastics, including their size, composition, abundance, and potential ecological impacts.

As research and technological advancements progress, it is expected that the field of microplastic detection will continue to evolve, leading to more efficient and reliable methods for monitoring and assessing the presence and impacts of microplastics in the environment.

REFERENCES

Agrawal, R., Kumar, R., Rai, S., Pathak, A. K., Rai, A. K., & Rai, G. K. (2011). LIBS: A quality control tool for food supplements. *Food Biophysics, 6*(4), 527–533. https://doi.org/10.1007/s11483-011-9235-y.

Andrady, A. L. (2011). Microplastics in the marine environment. *Marine Pollution Bulletin, 62*(8), 1596–1605. https://doi.org/10.1016/j.marpolbul.2011.05.030.

Asamoah, B. O., Kanyathare, B., Roussey, M., & Peiponen, K.-E. (2019). A prototype of a portable optical sensor for the detection of transparent and translucent microplastics in freshwater. *Chemosphere, 231*, 161–167. https://doi.org/10.1016/j.chemosphere.2019.05.114.

Asamoah, B. O., Roussey, M., & Peiponen, K.-E. (2020). On optical sensing of surface roughness of flat and curved microplastics in water. *Chemosphere, 254*, 126789. https://doi.org/10.1016/j.chemosphere.2020.126789.

Barrett, J., Chase, Z., Zhang, J., Holl, M. M. B., Willis, K., Williams, A., … Wilcox, C. (2020). Microplastic pollution in deep-sea sediments from the great Australian bight. *Frontiers in Marine Science, 7*. https://doi.org/10.3389/fmars.2020.576170.

Booth, A. M., Justynska, J., Kubowicz, S., Johnsen, H., & Frenzel, M. (2013). Influence of salinity, dissolved organic carbon and particle chemistry on the aggregation behaviour of methacrylate-based polymeric nanoparticles in aqueous environments. *International Journal of Environment and Pollution, 52*(1–2), 15–31. https://doi.org/10.1504/ijep.2013.056358.

Booth, A. M., Hansen, B. H., Frenzel, M., Johnsen, H., & Altin, D. (2016). Uptake and toxicity of methylmethacrylate-based nanoplastic particles in aquatic organisms. *Environmental Toxicology and Chemistry, 35*(7), 1641–1649. https://doi.org/10.1002/etc.3076.

Caputo, F., Vogel, R., Savage, J., Vella, G., Law, A., Della Camera, G., … Calzolai, L. (2021). Measuring particle size distribution and mass concentration of nanoplastics and microplastics: Addressing some analytical challenges in the sub-micron size range. *Journal of Colloid and Interface Science, 588*, 401–417. https://doi.org/10.1016/j.jcis.2020.12.039.

Carson, H. S., Nerheim, M. S., Carroll, K. A., & Eriksen, M. (2013). The plastic-associated microorganisms of the North Pacific Gyre. *Marine Pollution Bulletin, 75*(1), 126–132. https://doi.org/10.1016/j.marpolbul.2013.07.054.

Chen, D., Huang, Z., Wang, T., Ma, Y., Zhang, Y., Wang, G., & Zhang, P. (2020). High-throughput analysis of single particles by micro laser induced breakdown spectroscopy. *Analytica Chimica Acta, 1095*, 14–19. https://doi.org/10.1016/j.aca.2019.10.018.

Chen, D., Wang, T., Ma, Y., Wang, G., Kong, Q., Zhang, P., & Li, R. (2020). Rapid characterization of heavy metals in single microplastics by laser induced breakdown spectroscopy. *Science of the Total Environment, 743*, 140850. https://doi.org/10.1016/j.scitotenv.2020.140850.

Choobbari, M. L., Ciaccheri, L., Chalyan, T., Adinolfi, B., Thienpont, H., Meulebroeck, W., & Ottevaere, H. (2022). Batch analysis of microplastics in water using multi-angle static light scattering and chemometric methods. *Analytical Methods, 14*(39), 3840–3849. https://doi.org/10.1039/D2AY01215D.

Claessens, M., Meester, S. D., Landuyt, L. V., Clerck, K. D., & Janssen, C. R. (2011). Occurrence and distribution of microplastics in marine sediments along the Belgian coast. *Marine Pollution Bulletin, 62*(10), 2199–2204. https://doi.org/10.1016/j.marpolbul.2011.06.030.

Cole, M., Lindeque, P., Fileman, E., Halsband, C., Goodhead, R., Moger, J., & Galloway, T. S. (2013). Microplastic ingestion by zooplankton. *Environmental Science & Technology, 47*(12), 6646–6655. https://doi.org/10.1021/es400663f.

Cole, M., Lindeque, P., Halsband, C., & Galloway, T. S. (2011). Microplastics as contaminants in the marine environment: A review. *Marine Pollution Bulletin, 62*(12), 2588–2597. https://doi.org/10.1016/j.marpolbul.2011.09.025.

Croft, W. J. (2006). *Under the Microscope: A Brief History of Microscopy* (Vol. 5). World Scientific, Singapore.

da Costa, J. P., Santos, P. S. M., Duarte, A. C., & Rocha-Santos, T. (2016). (Nano)plastics in the environment – Sources, fates and effects. *Science of the Total Environment, 566–567*, 15–26. https://doi.org/10.1016/j.scitotenv.2016.05.041.

De Giacomo, A., Dell'Aglio, M., De Pascale, O., Longo, S., & Capitelli, M. (2007). Laser induced breakdown spectroscopy on meteorites. *Spectrochimica Acta Part B: Atomic Spectroscopy, 62*(12), 1606–1611. https://doi.org/10.1016/j.sab.2007.10.004.

Dick, W. D., Ziemann, P. J., & McMurry, P. H. (2007). Multiangle light-scattering measurements of refractive index of submicron atmospheric particles. *Aerosol Science and Technology, 41*(5), 549–569. https://doi.org/10.1080/02786820701272012.

Dockery, C. R., & Goode, S. R. (2003). Laser-induced breakdown spectroscopy for the detection of gunshot residues on the hands of a shooter. *Applied Optics, 42*(30), 6153–6158. https://doi.org/10.1364/AO.42.006153.

Duis, K., & Coors, A. (2016). Microplastics in the aquatic and terrestrial environment: Sources (with a specific focus on personal care products), fate and effects. *Environmental Sciences Europe, 28*(1), 1–25.

Egerton, R. F. (2005). *Physical Principles of Electron Microscopy* (Vol. 56). Springer, New York.

Enders, K., Lenz, R., Stedmon, C. A., & Nielsen, T. G. (2015). Abundance, size and polymer composition of marine microplastics ≥10μm in the Atlantic Ocean and their modelled vertical distribution. *Marine Pollution Bulletin, 100*(1), 70–81.https://doi.org/10.1016/j.marpolbul.2015.09.027.

Eriksen, M., Mason, S., Wilson, S., Box, C., Zellers, A., Edwards, W., … Amato, S. (2013). Microplastic pollution in the surface waters of the Laurentian Great Lakes. *Marine Pollution Bulletin, 77*(1), 177–182. https://doi.org/10.1016/j.marpolbul.2013.10.007.

Filella, M. (2015). Questions of size and numbers in environmental research on microplastics: Methodological and conceptual aspects. *Environmental Chemistry, 12*(5), 527–538. https://doi.org/10.1071/EN15012.

Fischer, D., Kaeppler, A., & Eichhorn, K. J. (2015). Identification of microplastics in the marine environment by Raman microspectroscopy and imaging. *American Laboratory, 47*(3), 32–34.

Foekema, E. M., De Gruijter, C., Mergia, M. T., van Franeker, J. A., Murk, A. J., & Koelmans, A. A. (2013). Plastic in North Sea Fish. *Environmental Science & Technology, 47*(15), 8818–8824. https://doi.org/10.1021/es400931b.

Fotopoulou, K. N., & Karapanagioti, H. K. (2012). Surface properties of beached plastic pellets. *Marine Environmental Research, 81*, 70–77. https://doi.org/10.1016/j.marenvres.2012.08.010.

Frias, J. P. G. L., Otero, V., & Sobral, P. (2014). Evidence of microplastics in samples of zooplankton from Portuguese coastal waters. *Marine Environmental Research, 95*, 89–95. https://doi.org/10.1016/j.marenvres.2014.01.001.

Fries, E., Dekiff, J. H., Willmeyer, J., Nuelle, M.-T., Ebert, M., & Remy, D. (2013). Identification of polymer types and additives in marine microplastic particles using pyrolysis-GC/MS and scanning electron microscopy. *Environmental Science: Processes & Impacts, 15*(10), 1949–1956. https://doi.org/10.1039/C3EM00214D.

Gaudiuso, R., Dell'Aglio, M., Pascale, O. D., Senesi, G. S., & Giacomo, A. D. (2010). Laser induced breakdown spectroscopy for elemental analysis in environmental, cultural heritage and space applications: A review of methods and results. *Sensors, 10*(8), 7434–7468.

Gies, S., Schömann, E.-M., Anna Prume, J., & Koch, M. (2020). Exploring the potential of time-resolved photoluminescence spectroscopy for the detection of plastics. *Applied Spectroscopy, 74*(9), 1161–1166.

Gigault, J., Pedrono, B., Maxit, B., & Ter Halle, A. (2016). Marine plastic litter: The unanalyzed nano-fraction. *Environmental Science: Nano, 3*(2), 346–350. https://doi.org/10.1039/C6EN00008H.

Goodhew, P. J., & Humphreys, J. (2000). *Electron Microscopy and Analysis*. CRC Press, Boca Raton, FL.

Guo, J., Lu, Y., Cheng, K., Song, J., Ye, W., Li, N., & Zheng, R. (2017). Development of a compact underwater laser-induced breakdown spectroscopy (LIBS) system and preliminary results in sea trials. *Applied Optics, 56*(29), 8196–8200. https://doi.org/10.1364/AO.56.008196.

Harrison, J. P., Ojeda, J. J., & Romero-González, M. E. (2012). The applicability of reflectance micro-Fourier-transform infrared spectroscopy for the detection of synthetic microplastics in marine sediments. *Science of the Total Environment, 416*, 455–463. https://doi.org/10.1016/j.scitotenv.2011.11.078.

Haynes, R. (2013). *Optical Microscopy of Materials*. Springer Science & Business Media, New York.

Hidalgo-Ruz, V., Gutow, L., Thompson, R. C., & Thiel, M. (2012). Microplastics in the marine environment: A review of the methods used for identification and quantification. *Environmental Science & Technology, 46*(6), 3060–3075. https://doi.org/10.1021/es2031505.

Hodgson, R. J. W. (2001). Genetic algorithm approach to the determination of particle size distributions from static light-scattering data. *Journal of Colloid and Interface Science, 240*(2), 412–418. https://doi.org/10.1006/jcis.2001.7652.

Huang, W., Yang, L., Yang, G., & Li, F. (2018). Microfluidic multi-angle laser scattering system for rapid and label-free detection of waterborne parasites. *Biomedical Optics Express, 9*(4), 1520–1530. https://doi.org/10.1364/BOE.9.001520.

Hyeon, Y., Kim, S., Ok, E., & Park, C. (2023). A fluid imaging flow cytometry for rapid characterization and realistic evaluation of microplastic fiber transport in ceramic membranes for laundry wastewater treatment. *Chemical Engineering Journal, 454*, 140028. https://doi.org/10.1016/j.cej.2022.140028.

Imhof, H. K., Ivleva, N. P., Schmid, J., Niessner, R., & Laforsch, C. (2013). Contamination of beach sediments of a subalpine lake with microplastic particles. *Current Biology, 23*(19), R867–R868. https://doi.org/10.1016/j.cub.2013.09.001.

Imhof, H. K., Laforsch, C., Wiesheu, A. C., Schmid, J., Anger, P. M., Niessner, R., & Ivleva, N. P. (2016). Pigments and plastic in limnetic ecosystems: A qualitative and quantitative study on microparticles of different size classes. *Water Research, 98*, 64–74. https://doi.org/10.1016/j.watres.2016.03.015.

Imhof, H. K., Schmid, J., Niessner, R., Ivleva, N. P., & Laforsch, C. (2012). A novel, highly efficient method for the separation and quantification of plastic particles in sediments of aquatic environments. *Limnology and Oceanography: Methods, 10*(7), 524–537. https://doi.org/10.4319/lom.2012.10.524.

Ivleva, N. P. (2021). Chemical analysis of microplastics and nanoplastics: Challenges, advanced methods, and perspectives. *Chemical Reviews, 121*(19), 11886-11936. https://doi.org/10.1021/acs.chemrev.1c00178.

Konde, S., Ornik, J., Prume, J. A., Taiber, J., & Koch, M. (2020). Exploring the potential of photoluminescence spectroscopy in combination with Nile Red staining for microplastic detection. *Marine Pollution Bulletin, 159*, 111475. https://doi.org/10.1016/j.marpolbul.2020.111475.

Lenz, R., Enders, K., Stedmon, C. A., Mackenzie, D. M. A., & Nielsen, T. G. (2015). A critical assessment of visual identification of marine microplastic using Raman spectroscopy for analysis improvement. *Marine Pollution Bulletin, 100*(1), 82–91. https://doi.org/10.1016/j.marpolbul.2015.09.026.

Löder, M. G. J., & Gerdts, G. (2015). Methodology used for the detection and identification of microplastics—A critical appraisal. In M. Bergmann, L. Gutow, & M. Klages (Eds.), *Marine Anthropogenic Litter* (pp. 201–227). Cham: Springer International Publishing.

Lusher, A. L., McHugh, M., & Thompson, R. C. (2013). Occurrence of microplastics in the gastrointestinal tract of pelagic and demersal fish from the English Channel. *Marine Pollution Bulletin, 67*(1), 94–99. https://doi.org/10.1016/j.marpolbul.2012.11.028.

Maes, T., Jessop, R., Wellner, N., Haupt, K., & Mayes, A. G. (2017). A rapid-screening approach to detect and quantify microplastics based on fluorescent tagging with Nile Red. *Scientific Reports, 7*(1), 44501. https://doi.org/10.1038/srep44501.

Napper, I. E., & Thompson, R. C. (2016). Release of synthetic microplastic plastic fibres from domestic washing machines: Effects of fabric type and washing conditions. *Marine Pollution Bulletin, 112*(1), 39–45. https://doi.org/10.1016/j.marpolbul.2016.09.025.

Phuong, N. N., Zalouk-Vergnoux, A., Poirier, L., Kamari, A., Châtel, A., Mouneyrac, C., & Lagarde, F. (2016). Is there any consistency between the microplastics found in the field and those used in laboratory experiments? *Environmental Pollution, 211*, 111–123. https://doi.org/10.1016/j.envpol.2015.12.035.

Pirc, U., Vidmar, M., Mozer, A., Kržan, A. J. E. S., & Research, P. (2016). Emissions of microplastic fibers from microfiber fleece during domestic washing. *Environmental Science and Pollution Research, 23*, 22206–22211.

Reddy, M. S., Shaik, B., Adimurthy, S., & Ramachandraiah, G. (2006). Description of the small plastics fragments in marine sediments along the Alang-Sosiya ship-breaking yard, India. *Estuarine, Coastal and Shelf Science, 68*(3), 656–660. https://doi.org/10.1016/j.ecss.2006.03.018.

Reimer, L. J. M. S., & Technology. (2000). Scanning electron microscopy: Physics of image formation and microanalysis. *Measurement Science and Technology, 11*(12), 1826–1826.

Rios Mendoza, L. M., & Jones, P. R. (2015). Characterisation of microplastics and toxic chemicals extracted from microplastic samples from the North Pacific Gyre. *Environmental Chemistry, 12*(5), 611–617. https://doi.org/10.1071/EN14236.

Rocha-Santos, T., & Duarte, A. C. (2015). A critical overview of the analytical approaches to the occurrence, the fate and the behavior of microplastics in the environment. *TrAC Trends in Analytical Chemistry, 65*, 47–53. https://doi.org/10.1016/j.trac.2014.10.011.

Schmid, T., Opilik, L., Blum, C., & Zenobi, R. (2013). Nanoscale chemical imaging using tip-enhanced Raman spectroscopy: A critical review. *Angewandte Chemie International Edition, 52*(23), 5940–5954. https://doi.org/10.1002/anie.201203849.

Sharma, N., Chi, C.-H., Swaminathan, N., Dabur, D., & Wu, H.-F. (2021). Introducing Stanene oxyboride nanosheets as white light emitting probe for selectively identifying <5 µm microplastic pollutants. *Sensors and Actuators B: Chemical, 348*, 130617. https://doi.org/10.1016/j.snb.2021.130617.

Shaw, D. G., & Day, R. H. (1994). Colour- and form-dependent loss of plastic micro-debris from the North Pacific Ocean. *Marine Pollution Bulletin, 28*(1), 39–43. https://doi.org/10.1016/0025-326X(94)90184-8.

Smith, R., Wright, K. L., & Ashton, L. (2016). Raman spectroscopy: An evolving technique for live cell studies. *Analyst, 141*(12), 3590–3600. https://doi.org/10.1039/C6AN00152A.

Sommer, C., Schneider, L. M., Nguyen, J., Prume, J. A., Lautze, K., & Koch, M. (2021). Identifying microplastic litter with Laser Induced Breakdown Spectroscopy: A first approach. *Marine Pollution Bulletin, 171*, 112789. https://doi.org/10.1016/j.marpolbul.2021.112789.

Tagg, A. S., Sapp, M., Harrison, J. P., & Ojeda, J. J. (2015). Identification and quantification of microplastics in wastewater using focal plane array-based reflectance micro-FT-IR imaging. *Analytical Chemistry, 87*(12), 6032–6040. https://doi.org/10.1021/acs.analchem.5b00495.

Török, P., & Kao, F.-J. (2007). *Optical Imaging and Microscopy: Techniques and Advanced Systems* (Vol. 87): Springer, Berlin.

Van Cauwenberghe, L., Claessens, M., Vandegehuchte, M. B., & Janssen, C. R. (2015). Microplastics are taken up by mussels (Mytilus edulis) and lugworms (Arenicola marina) living in natural habitats. *Environmental Pollution, 199*, 10–17. https://doi.org/10.1016/j.envpol.2015.01.008.

Van Cauwenberghe, L., Vanreusel, A., Mees, J., & Janssen, C. R. (2013). Microplastic pollution in deep-sea sediments. *Environmental Pollution, 182*, 495–499. https://doi.org/10.1016/j.envpol.2013.08.013.

Watts, A. J. R., Lewis, C., Goodhead, R. M., Beckett, S. J., Moger, J., Tyler, C. R., & Galloway, T. S. (2014). Uptake and retention of microplastics by the shore crab Carcinus Maenas. *Environmental Science & Technology, 48*(15), 8823–8830. https://doi.org/10.1021/es501090e.

Weinstein, J. E., Crocker, B. K., & Gray, A. D. (2016). From macroplastic to microplastic: Degradation of high-density polyethylene, polypropylene, and polystyrene in a salt marsh habitat. *Environmental Toxicology and Chemistry, 35*(7), 1632–1640. https://doi.org/10.1002/etc.3432.

Yang, L., Qiao, F., Lei, K., Li, H., Kang, Y., Cui, S., & An, L. (2019). Microfiber release from different fabrics during washing. *Environmental Pollution, 249*, 136–143. https://doi.org/10.1016/j.envpol.2019.03.011.

Zettler, E. R., Mincer, T. J., & Amaral-Zettler, L. A. (2013). Life in the "Plastisphere": Microbial communities on plastic marine debris. *Environmental Science & Technology, 47*(13), 7137–7146. https://doi.org/10.1021/es401288x.

Zhang, K., Su, J., Xiong, X., Wu, X., Wu, C., & Liu, J. (2016). Microplastic pollution of lakeshore sediments from remote lakes in Tibet plateau, China. *Environmental Pollution, 219*, 450–455. https://doi.org/10.1016/j.envpol.2016.05.048.

Zhao, S., Zhu, L., & Li, D. (2015). Microplastic in three urban estuaries, China. *Environmental Pollution, 206*, 597–604. https://doi.org/10.1016/j.envpol.2015.08.027.

3 Plastic Peril
Unraveling Microplastics Threats to Health and Ecosystems

Ajay Valiyaveettil Salimkumar, Mary Carolin Kurisingal Cleetus, Dorcas Akua Essel, Muhammad Arslan Aslam, Asadulla Hil Galib, Prabhakaran M. P., and Rejish Kumar V. J.

3.1 INTRODUCTION

When Bakelite, the first synthetic plastic, was discovered in 1907, it transformed the study of polymers and contemporary life by bringing a variety of polymers and plastic formulas into use daily (Mossman 2008). However, the fast increase in global plastic manufacture did not occur until the 1950s (Klöckner et al. 2021). Over 380 million tonnes of plastics are produced annually in the world due to their promise, affordability, and appropriateness for a wide range of commercial and industrial uses (Plastics Europe Publications 2020). However, due to its widespread dispersion in freshwater and marine habitats, what was once and is still considered a revolutionary substance has gradually evolved into a global environmental hazard (Zeng et al. 2018). A new class of plastic pollutants known as microplastics (MPs) has emerged because of the modifications that plastics undergo in the environment (Koyilath et al. 2022). The term "microplastics" was first used by Thompson et al. (2004) to characterize the buildup of minute particles of plastic in marine sediments and the water column of European seas. Plastic particles under 5 mm are often referred to as microplastics or MPs (Thompson 2015). The usage of the term "often" revealed that there is not agreement on its proper definition. According to Frias and Nash (2019), the acceptable size range for the upper limit is 500 µm to either 1 mm or 5 mm, and the lower limit is 1–20 µm. Cole et al. (2011) further clarified the concept of MPs by classifying them as primary (made to be of microscopic dimensions) or secondary (coming from degradation and fragmentation processes in the environment). Frias and Nash (2019) proposed all-inclusive definitions of MPs, which are as follows: "MPs are any synthetic solid particle or polymeric matrix, with regular or irregular shape and with size ranging from 1 µm to 5 mm, of either primary or secondary manufacturing origin, which are insoluble in water."

DOI: 10.1201/9781003438793-3

There is agreement that it is critical to investigate MPs in natural habitats because of the dangers that can pose to the environment; however, there is a lack of consensus on the proper definition of MPs (Koyilath et al. 2022). MPs, sometimes referred to as primary MPs, can be produced directly and are found in a variety of personal care and cosmetic products (PPCPs) (Van Wezel et al. 2016). Additionally, they can be created by the secondary MPs process, which is the erosion of big plastic waste through exposure to environmental stressors such as water, wind, and sunshine (Galgani et al. 2013). Microplastics can be discharged into the environment from a variety of sources. The top MP environmental contributions are from the personal care and cosmetics industries (Anderson et al. 2016). The majority of exfoliants and cleansers employ MPs as a replacement for commonly used ingredients including almonds, oats, and maize (Kristanti et al. 2023). According to the estimates, each application of an exfoliant can release up to 94,000 MPs (Anderson et al. 2016).

Polyester, the most common textile fiber, which can replace cotton at a lower cost, plays a significant role in the persistence of MPs in terrestrial, avian, and aquatic environments (Carney Almroth et al. 2018). The physical and chemical stresses that textiles go through throughout the washing process will result in the release of microfibers (De Falco et al. 2019). A former study found that the amount of fiber released per kilogram of washing might reach over 110,000 fibers (Carney Almroth et al. 2018; De Falco et al. 2019). Global average use of garment fibers has reached 80% over the past 20 years, primarily driven by a substantial 300% increase in the usage of synthetic fibers (Carney Almroth et al. 2018). In 2011, China's agricultural products used more than 1.20 million tons of plastic components. Contrarily, sewage sludge used as fertilizer in agriculture had MPs of around 250,000 and 170,000 tons in Europe and America, respectively (Ding et al. 2021).

According to a recent study, wastewater treatment facilities (WWTPs) may be a significant source in the release of MPs into the environment (Browne et al. 2011). The production and disposal of MPs in the home and commercial wastewater streams, as well as their frequent entry into the stormwater drainage systems, make WWTPs a point source of MPs (Koyilath et al. 2022). WWTPs are a significant source of MPs in the environment. MPs from wastewater treatment plants can be inefficiently treated and introduced into marine and terrestrial environments, posing potential risks to ecosystems and biota (Conley et al. 2019). MPs emitted from these plants may pose unique toxicological risks due to their immediate bioaccessibility (Hurley et al. 2018). In terms of ecotoxicological impacts, MPs have been reported to cause physical and photosynthetic changes in *Chlorella pyrenoidosa* (Mao et al. 2018) and damaged cell walls (Liu et al. 2019). This chapter explains the health risks and ecotoxicological impacts of MPs, highlighting the fate of transport mechanisms and the current policy measures adopted in different countries. It also identifies the challenges and drawbacks of these policy measures.

3.2 HEALTH RISKS OF MPS

3.2.1 Ingestion and Accumulation

The most frequent way that marine animals and MPs interact is through ingestion. Plastic breaks down into numerous sizes and segments in the aquatic environment. According to Pandey et al. (2022), these segments may stay floating in the water column, or they may combine with other organic components and sink to the bottom. They are mistakenly eaten as food because of their similarity with planktons (Ding et al. 2021), or they are absorbed through filtering processes during feeding in the water column. Thus, the primary uptake pathway of MPs in most aquatic organisms is through the gastrointestinal tract and gills. It has been estimated that MPs have been reported in about 100 marine species through ingestion (Kershaw & Rochman 2015), and more than half of these species are eaten as food by humans. Furthermore, plastic debris or MPs have been detected in the digestive tracts or bodily tissues of many marine organisms (Yu et al. 2020). Other reports show MPs can accumulate in large amounts in zooplankton, amphipods, mussels, and several fishes (Cole et al. 2015; Li et al. 2020).

3.2.2 Bioaccumulation and Biomagnification

Biomagnification and bioaccumulation are used to describe the concentration of the notified chemical concentration in the ecological food web. Bioaccumulation occurs when the uptake of a contaminant through ingestion or respiration from the environment, in this case, MP, is greater than the ability of an organism to egest the contaminant (Miller et al. 2020). However, when the concentration of MP in one organism is higher than its prey across the food chain, then it gets biomagnified. MPs are incredibly mobile and ubiquitous in the aquatic environment due to their lightweight and insolubility (Hurley et al. 2018) and are considered highly bioavailable contaminants to marine organisms and the organisms that consume them. These contaminants can be transferred from prey to predators through indirect trophic transfer by means of ingestion. Humans may experience the greatest exposure as a result of the possible biomagnification of these marine contaminants.

3.2.3 Possible Effects on Organ Systems

Although they could induce chronic toxicity, which is observed as a major problem with long-term exposure, MPs do not immediately kill living things. Chemicals adsorbed on the surface of MPs and additives are the principal sources of toxicity. These chemical substances enter organisms with MP particles and may be released during desorption processes, harming the organism (Zimmermann et al. 2020). Recent research discovered that the shape and amount of accumulated MPs in distinct marine species' organs vary greatly (Campanale et al. 2020). According to Zimmermann et al. (2020), the polymer components used to produce plastic products could be the principal source of toxicity. Second, because of their small sizes, MPs can injure organisms and can cause inflammation. Some animals may suffer from malnutrition and alterations in reproduction because of swallowing minute MPs (Jewett et al. 2022). MP exposure alters the genes, cells, tissues, and organs of marine animals at many levels through a variety of processes. However, most studies in the field focus on the potential effects on species in the Mollusca Phyla (class Bivalvia) due to their ecological importance as filter feeders (Li et al. 2016). Changes in these organisms owing to MP buildup cause harm to their neurological and digestive systems, affecting their health. Cole et al. (2015) similarly observed that the presence of MPs resulted in a significant reduction in the food intake of copepods, indicating negative effects on the zooplankton's health.

3.2.4 Potential Implications for Human Health

The possibility of MPs infiltrating the marine food chain through biomagnification raises concerns about food safety and its implications for human health. Humans are exposed to MPs through mouth ingestion and nose inhalation (Cox et al. 2019). Studies have shown the presence of different sizes of polyethylene terephthalate, polypropylene, and polyethylene in products such as salt and drinking water globally (Wright et al. 2013; Zhang et al. 2020). These MPs enter the gastrointestinal tract and may cause tissue inflammation and organ damage or enter the circulatory system and accumulate in the liver. Due to their size, fibers from clothing can travel short and long distances in the atmosphere, are absorbed by humans (Dris et al. 2017), and can harm the body's respiratory system when inhaled. The absence of data on human exposure is a key concern for establishing the dangers of MPs to human health (Cox et al. 2019), and thus, further research is needed to determine the extent of harm that can be caused by MPs in humans. Moreover, with the goal of establishing a biological foundation for human risk assessment, it is important to revise technical strategies that are involved in wastewater treatments to reduce the introduction of MP into the aquatic environment.

3.3 ECOTOXICOLOGICAL IMPACTS OF MPS

3.3.1 Effects on Aquatic Organisms

Sewage effluents have been identified as one of the significant sources of MPs in the aquatic environment, and ingestion by the biota is one of the potential sinks. Although researchers have reported the ingestion and accumulation of MPs in marine vertebrates and invertebrates of pelagic to sedimentary zones of diverse marine species, from plankton to the higher trophic levels (Parolini et al. 2023), including the deep-sea sediments (Van Cauwenberghe et al. 2013), and its accumulation in tissues of organisms. However, the effects caused by these need to be studied making it one of the foremost concerns. Reduced photosynthetic activity, in addition to the physical damage such as unclear pyrenoids, distorted thylakoids, damaged cell membrane, and oxidative stress was noticed in the algal species *Chlorella pyrenoidosa* in response to different size polystyrene MPs exposure under dose-dependent conditions (Mao et al. 2018). Liu et al. (2019) observed a similar finding that smaller MPs damaged the cell wall by adhering to the algal surface. In contrast, larger MPs had adverse effects by obstructing light transfer and impairing photosynthesis in *Scenedesmus obliquus*. In compliance with Cole et al. (2016), MPs have the potential to drastically change the density, structural integrity, and sinking rates of fecal pellets ejected by marine zooplankton. MPs obstruct the food intake capacity of zooplankton through intestinal obstructions and prolonged gut retention periods of plastics (Cole et al. 2013). Furthermore, a significant decrease in fecundity (Lee et al. 2013), relatively slower growth rates (Lo et al. 2018), and an increased mortality rate (Lee et al. 2013) were also observed. MPs can also have unfavorable effects on the echinoderm *Holothuria leucospilota,* by its accumulation in the gut, respiratory tree, and tentacles (Ahmed et al. 2023). Altered gene expression by microfibres in *Apostichopus japonicus* (Mohsen et al. 2021), accumulation of microfibres in the gut and coelomic fluid of *Holothuria cinerascens* (Iwalaye et al. 2020) were also observed. Various effects have been recorded in the crustaceans such as inhibiting growth, disturbed lipid metabolism, and decreased fatty acid metabolism in *Cherax quadricarinatus* (Chen et al. 2020), altering the intestinal microbiota, metabolite molecules hemolymph protein profiles in *Litopenaeus vannamei* (Duan et al. 2021). Cellular and subcellular level effects have been observed including histological alterations and strong inflammatory responses and lysosomal membrane destabilization when exposed to a longer exposure time in *Mytilus edulis* (Von Moos et al. 2012), malformations and developmental defects in *Crassostrea gigas* (Bringer et al. 2020). In general, filter feeders and deposit feeders may be more susceptible to MPs and more readily bioavailable. Fish exposed to MPs can develop tissue damage, oxidative stress, alterations in the expression of immune-related genes, and impaired levels of antioxidants. Moreover, it exhibits aberrant behavior, decreased growth, and neurotoxicity (Bhuyan 2022).

3.3.2 Effects on Terrestrial Organisms

Despite significant methodological shortcomings, there has been ample research in the aquatic environment on identifying MPs. Compared to aquatic ecosystems, less research has been done on fate, transfer, and impacts of MPs in complex terrestrial ecosystems. The physiochemistry and biota of the soil have a direct effect on the fate of MPs (He et al. 2018). The MP that ends up in terrestrial soil may originate from the sewage plant sludge or from farming activities. There have been relatively less studies available on the effects of MPs on soil organisms. MPs can be uptake by plant roots and can impair growth and performance by obstructing pores or light, deteriorating roots mechanically, impeding gene expression, and secreting additives. They can also affect soil properties, soil beings, and the bioavailability of other pollutants (Li et al. 2022a). The investigation by Jiang et al. (2020) has concluded that polystyrene MPs can cause intestinal cell disintegration, DNA damage, oxidative stress, and other histopathological abnormalities in earthworm *Eisenia fetida*. Besides, negative effects on reproduction have been observed in the nematode *Caenorhabditis elegans* exposed to MPs. Smaller-sized nylon MPs hindered the reproduction of the Enchytraeid, *Enchytraeus crypticus*, without affecting its survival (Lahive et al. 2019). However, the research by Kokalj et al. (2018) found that the MP from plastic bags and facial cleansers had no impact on the feeding rate, body mass, or energy stores in the digestive glands of the terrestrial isopod *Porcellio scaber*. Likewise, mealworms, enchytraeids, and woodlice are not impacted by exposure to MPs made from disposable surgical masks. However, it led to alterations in the energy-related indicators in mealworm larvae and a temporary immunological response of increased hemocyte count in woodlice (Kokalj et al. 2022). Meanwhile, Abdel-Zaher et al. (2023) have demonstrated that MPs have damaging effects on the red blood cells of mice and may modify the composition and wide range of intestinal microbes (Li et al. 2020). MP ingestion in chicken can result in intestinal metabolic disorders, causing gut microbial dysbiosis as well as lowering growth parameters and antioxidant ability (Li et al. 2023).

3.3.3 Impacts on Ecosystem Services

MPs pose a serious threat to the aquatic ecosystem since they will severely impact the water quality, which is essential to most life on earth, including humans. The chemicals in the plastics and MPs may leach into the environment they are exposed to and cause substantial concerns for the entire ecosystem. According to reports, these leached additives provide a greater environmental risk than plastic themselves (da Costa et al. 2022) in contaminating the water. This contamination will likely affect both aquatic life that depends on clean water for survival and the availability of clean water for humans. Moreover, sediment's physicochemical characteristics, enzymatic activity, and diversity of microbial communities can all be affected by the accumulation of MPs in the sediment (Li et al. 2022b). Besides, MPs may affect marine phytoplankton and zooplankton growth and reproduction, which may impact ocean carbon sequestration (Shen et al. 2020). MPs ingested by the organisms cause detrimental effects to the organisms themselves, transmitting to the food chain and ultimately disrupting the functioning of the entire biodiversity. Higher trophic levels, including commercially important fish species, can be affected by the bioaccumulation of MPs through the food chain transfer and ultimately become a threat to human health. The consequences can differ between organisms, and they can be very severe in numerous populations, which can cause a loss in biodiversity. It's critical that our knowledge of how MPs affect ecosystem services remains limited, and more research is necessary to understand the severity of the problem.

3.3.4 Implications for Food Webs

MPs have been ingested by most terrestrial and aquatic creatures, including human beings, and their possible fate, means of transmission, and effects are raising concerns. Since MPs can be bioavailable, they can penetrate the complex food web. Primary producers, the building blocks of a trophic chain or web, consume MPs in terrestrial and aquatic habitats and can have a greater effect on the entire ecosystem. By limited primary output, energy flow between food chains may be affected which can lead to a decrease in species abundance in the higher trophic levels. Based on the previously reviewed data, Figure 3.1 depicts the possible flow of MPs within the food web.

3.4 CURRENT REGULATIONS AND POLICY

There has been increasing concern among the public, media, decision-makers, and the scientific community regarding potential risks to human health and the environment caused by the widespread use of plastics and the subsequent problem of plastic pollution on a global scale (Koelmans et al. 2017). As a result, several legislative proposals have been adopted to fight plastic pollution on a national and worldwide scale (Table 3.1). To solve the challenges of microplastics (MP) contamination, researchers and other members of the plastic value chain must collaborate in a coordinated, multi-sectoral effort (Worm et al. 2017). The legislation largely relied on prohibitions, the application of taxes, and volunteer efforts to reduce and reuse of plastics (Lam et al. 2018).

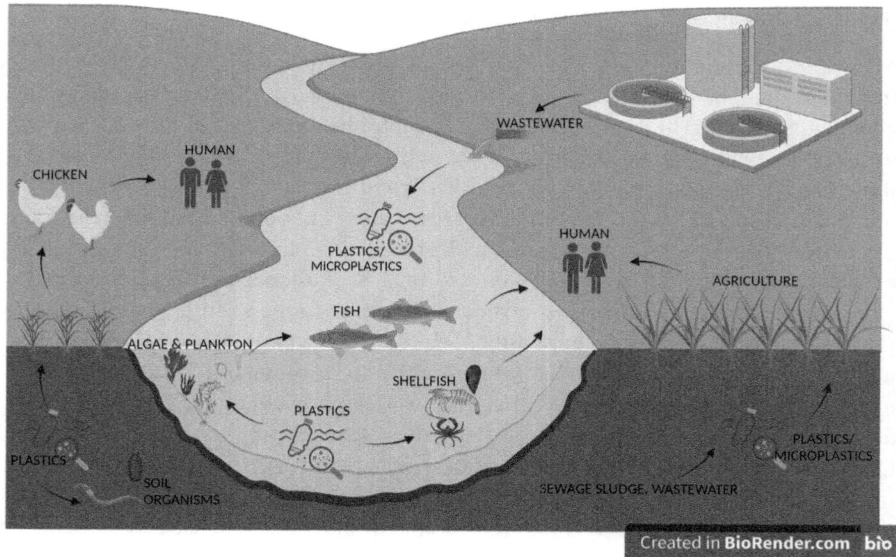

FIGURE 3.1 The basic representation of the possible food web transfer of MPs in terrestrial and aquatic environments.

TABLE 3.1
Overview of Policy Initiatives Implemented at the National Level

Policy Measures	Countries	Measure	References
Partial or complete ban on single-use plastic	34 African states	Plastic is banned	DAWE (2020), Bezerra et al. (2021), Cloché-Dubois et al. (2021), NCSL (2021)
	France	Ban and recycling of plastic	
	Australia	Ban on the plastic export	
	Hawaii	Banned plastic bags at shops	
	California	Single-use plastic bags forbidden	
	Vermont	Partially banned plastic bags	
Taxes or Levies	Sweden	Tax on Plastic manufacture and import	MESTECC (2019), Smith (2020), Sweden (2020), Bezerra et al. (2021), Ernst and Young (2021)
	Angola	Tax on the use of plastic bags	
	Botswana		
	Lesotho		
	South Africa		
	Italy	Tax on the plastic industry	
	Malaysia	Fees on the use of plastic bags	
	England	Plastic packaging tax	

An entire or partial ban on single-use plastics is the often-used policy measure, and 34 African countries have already done so (Bezerra et al. 2021). Single-use plastics are prohibited in France, and circular economic models are encouraged

(Cloché-Dubois et al. 2021). Additionally, Australia forbade the export of plastic and developed a flowchart for trash management and recycling (DAWE 2020). Vermont has added more limits on the containers of single-use straws and polystyrene, joining Oregon, Maine, Delaware, Vermont, and Connecticut in outlawing single-use plastic bags (NCSL 2021). Single-use plastic bags at grocery shops and other businesses are prohibited in Hawaii (NCSL 2021). California has enacted a statewide ban on all single-use plastic bags and a 10-cent minimum fee for recycled paper (NCSL 2021).

Taxes and levies on plastic items are the often-used measures for reducing plastic consumption globally. To stop the spread of MPs and achieve the European Union's (EU) objective of having less than 40 plastic bags used per person yearly by 2025, a tax has been imposed on makers and importers of plastic bags in Sweden (2020). Angola, Botswana, Lesotho, South Africa, and four other members of the Southern African Development Community charge fees for plastic bags (Bezerra et al. 2021). The Italian Plastic Packaging Law levies taxes on single-use plastics, plastic producers, commercial consumers, and commercial dealers of plastics (Ernst & Young 2021). Malaysia established a communication, education, and public awareness unit in addition to imposing a tax on single-use plastic bags and plastic producers. Additionally, it promotes the development of plastic substitutes like bio bags (MESTECC 2019). There is a plastics packaging tax in England as well (Smith 2020).

The Microbead-Free Waters Act (2005), which was created and went into effect in the United States in 2017, was primarily created to address MP pollution and, by extension, that of plastic bags. It prohibits the sale of cosmetics using microbeads and encourages the use of plastic substitutes such as biodegradable materials. China has plastic pollution regulations designed to control plastic pollution and provides plastic alternatives that forbid the disposal of plastics (Zhang et al. 2018).

It was to make sure that by 2025, every plastic package sold could be recycled, reused, or composted (Smith 2020). With instruments at several phases of their lifecycle, including manufacturing, import, sale, use, and disposal, the Canadian Environmental Protection Act of 1999 (CEPA) was designed to fight plastic pollution (Environment and Climate Change 2020). The goal of the Korea Plastic Trash Control Plan was to recycle 70% of the plastic waste produced while lowering it by 50%. It outlines strategies to reestablish manufacturing and consumption mechanisms as well as a circular economy (Shin et al. 2020).

3.5 CHALLENGES AND DRAWBACKS

While the policies and initiatives mentioned above are aimed at addressing plastic pollution and promoting sustainable practices, they also face certain drawbacks and challenges. Different areas and nations' regulations regarding plastic pollution are neither uniform or consistent. Confusion and inefficiency can result from this, especially for companies that operate in several different jurisdictions. Regulation inconsistencies can also lead to the relocation of plastic manufacturing or waste disposal operations to regions with laxer restrictions, a situation known as "regulatory arbitrage." For instance, in France, MPs are permitted in areas with pharmaceuticals for human and veterinary use (Cloché-Dubois et al. 2021). Microbeads and plastic bags in personal care products are still legal in China (Zhang et al. 2018). Compared to other nations in Europe and the rest of the world, Sweden's plastic law exempts the import of bags less than 40 intended for personal use as well as bags intended for continuous use (Sweden 2020).

It might be difficult to strike a balance between economic concerns and environmental goals. Numerous sectors heavily rely on the manufacturing and use of plastic, and abrupt legislative changes may have negative economic effects, especially for companies and employees reliant on the plastic value chain. For instance, because of the dependency of the poor on the plastic industries, the plastic ban law in Africa encounters opposition from significant parties, weak implementation, and a lack of alternatives (Nyathi et al. 2020). Despite the talks, plastic manufacturers in Kenya and Uganda objected to the prohibition, citing the possibility of lost jobs and revenue (Behuria 2021). Plastic banning has not been successful in the majority of nations due to the financial power of plastic makers, distributors, and retailers, the lack of policy execution, the short time between the announcement of the ban and its effective date, and the inconsistent nature of national rules (Adam et al. 2020).

Unintended effects may result from some measures, such as outright bans on specific plastic goods. For instance, a restriction on single-use plastic bags may lead to a rise in the demand for alternative materials, which may be challenging to supply. Lack of national policies makes it difficult for domestic companies to flourish or for importers of biodegradable plastic to satisfy the demand (Oyake-Ombis et al. 2015). So, for innovation to be a useful instrument for reducing MP pollution, legislative support is necessary (Amankwa et al. 2021).

A policy strategy that tries to limit the usage of plastic bags and motivate the use of reusable alternatives is the implementation of fees or levies on them. Although this strategy has shown some promise in decreasing the use of plastic bags, it is not without difficulties and disadvantages. Taxes and levies are frequently ineffective because importers, producers, and retailers eventually pass the cost on to consumers (Adam et al. 2020). Prior research in South Africa (O'Brien & Thondhlana 2019) and Botswana (Madigele et al. 2017) revealed that those with university education and low incomes were less inclined to pay bag levies. High levies on bags could function as a reliable deterrent for excessive use of plastic bags. South Africa's usage of plastic bags considerably declined with the introduction of consumer plastic levies, even if it steadily climbed over time (Dikgang et al. 2012a). Given that plastic bags were provided without any fee, the transitory fall in plastic bag usage was likely caused by loss aversion on the part of the consumer (Dikgang et al. 2012b). However, it was not accompanied by user education and stakeholder engagements before execution, as was the case in Ireland. The plastic bag levy in South Africa did not succeed in curbing the overuse of plastic (Muposhi et al. 2021).

Addressing these challenges requires ongoing collaboration and continuous evaluation of policies to ensure they are effective, adaptable, and sustainable in the long term. It is important to engage with stakeholders from different sectors and learn from the experiences of regions/countries that have implemented successful policies to improve future approaches to tackle plastic pollution.

3.6 CONCLUSIONS

MPs limit the efficiency of wastewater treatment systems. Although wastewater systems are meant to remove as much plastic as possible from sewage, there is mounting evidence that not all MPs are removed from wastewater and are released back into the environment, where they eventually become ecologically harmful. There is an urgent need to examine water treatment systems and processes used to remove MPs from wastewater before they are released into the environment. Once in the environment, MPs undergo breakdown into different sizes and shapes that are mistakenly ingested as food by some organisms and will eventually travel to the higher trophic levels through the food web. Thus, it's critical to assess how various plastic particle sizes discharged will affect the species that consume them. As humans, we become vulnerable to contamination by MPs since we are the top predators. Even though research on the effects of plastics on human health is available, it will be interesting to investigate the potential routes by which MPs can enter into our bodies as well as the various toxicophysiological effects caused by different types of MPs.

REFERENCES

Abdel-Zaher, Souzan, Mahmoud S. Mohamed, and Alaa El-Din H. Sayed. "Hemotoxic effects of polyethylene microplastics on mice." *Frontiers in Physiology* 14 (2023): 1072797.

Adam, Issahaku, et al. "Policies to reduce single-use plastic marine pollution in West Africa." *Marine Policy* 116 (2020): 103928.

Ahmed, Quratulan, et al. "Analysis of Microplastic in Holothuria leucospilota (Echinodermata-Holothuroidea) and Sediments from Karachi coast, (Northern Arabian Sea)." *International Journal of Environment and Geoinformatics* 10.1 (2023): 161–169.

Amankwa, M. Opoku, et al. "The production of valuable products and fuel from plastic waste in Africa." *Discover Sustainability* 2 (2021): 1–11.

Anderson, A. G., et al. "Microplastics in personal care products: Exploring perceptions of environmentalists, beauticians and students." *Marine Pollution Bulletin* 113.1–2 (2016): 454–460.

Behuria, Pritish. "Ban the (plastic) bag? Explaining variation in the implementation of plastic bag bans in Rwanda, Kenya and Uganda." *Environment and Planning C: Politics and Space* 39.8 (2021): 1791–1808.

Bezerra, Joana Carlos, et al. "Single-use plastic bag policies in the Southern African development community." *Environmental Challenges* 3 (2021): 100029.

Bhuyan, Md Simul. "Effects of microplastics on fish and in human health." *Frontiers in Environmental Science* 10 (2022): 250.

Bringer, Arno, et al. "Experimental ingestion of fluorescent microplastics by pacific oysters, Crassostrea gigas, and their effects on the behaviour and development at early stages." *Chemosphere* 254 (2020): 126793.

Browne, Mark Anthony, et al. "Accumulation of microplastic on shorelines woldwide: Sources and sinks." *Environmental Science & Technology* 45.21 (2011): 9175–9179.

Campanale, Claudia, et al. "A detailed review study on potential effects of microplastics and additives of concern on human health." *International Journal of Environmental Research and Public Health* 17.4 (2020): 1212.

Carney Almroth, Bethanie M., et al. "Quantifying shedding of synthetic fibers from textiles; a source of microplastics released into the environment." *Environmental Science and Pollution Research* 25 (2018): 1191–1199.

Chen, Qiang, et al. "Effects of exposure to waterborne polystyrene microspheres on lipid metabolism in the hepatopancreas of juvenile redclaw crayfish, Cherax quadricarinatus." *Aquatic Toxicology* 224 (2020): 105497.

Cloché-Dubois C., et al. "CMS Expert Guide to plastics and packaging laws A snapshot of legal developments Plastics and packaging laws in France." *C. Law. Tax. Future* 6 (2021): 1–9.

Cole, Matthew, et al. "Microplastic ingestion by zooplankton." *Environmental Science & Technology* 47.12 (2013): 6646–6655.

Cole, Matthew, et al. "Microplastics alter the properties and sinking rates of zooplankton faecal pellets." *Environmental Science & Technology* 50.6 (2016): 3239–3246.

Cole, Matthew, et al. "Microplastics as contaminants in the marine environment: A review." *Marine Pollution Bulletin* 62.12 (2011): 2588–2597.

Cole, Matthew, et al. "The impact of polystyrene microplastics on feeding, function and fecundity in the marine copepod Calanus helgolandicus." *Environmental Science & Technology* 49.2 (2015): 1130–1137.

Conley, Kenda, et al. "Wastewater treatment plants as a source of microplastics to an urban estuary: Removal efficiencies and loading per capita over one year." *Water Research X* 3 (2019): 100030.

Cox, Kieran D., et al. "Human consumption of microplastics." *Environmental Science & Technology* 53.12 (2019): 7068–7074.

da Costa, Joao Pinto, et al. "Plastic additives and microplastics as emerging contaminants: Mechanisms and analytical assessment." *TrAC Trends in Analytical Chemistry* 158 (2022): 116898.

DAWE. (2020). "Australia Recycling and Waste Reduction Bill." Available online: https://www.aph.gov.au/Parliamentary_Business/Bills_Legislation/Bills_Search_Results/Result?bId=r6573 (accessed on 28 May 2023).

De Falco, Francesca, et al. "The contribution of washing processes of synthetic clothes to microplastic pollution." *Scientific Reports* 9.1 (2019): 6633.

Dikgang, Johane, Anthony Leiman, and Martine Visser. "Analysis of the plastic-bag levy in South Africa." *Resources, Conservation and Recycling* 66 (2012a): 59–65.

Dikgang, Johane, Anthony Leiman, and Martine Visser. "Elasticity of demand, price and time: Lessons from South Africa's plastic-bag levy." *Applied Economics* 44.26 (2012b): 3339–3342.

Ding, Runrun, Ling Tong, and Weicheng Zhang. "Microplastics in freshwater environments: Sources, fates and toxicity." *Water, Air, & Soil Pollution* 232 (2021): 1–19.

Dris, Rachid, et al. "A first overview of textile fibers, including microplastics, in indoor and outdoor environments." *Environmental Pollution* 221 (2017): 453–458.

Duan, Yafei, et al. "Toxicological effects of microplastics in Litopenaeus vannamei as indicated by an integrated microbiome, proteomic and metabolomic approach." *Science of the Total Environment* 761 (2021): 143311.

Environment and Climate Change Canada. "Canada One-Step Closer to Zero Plastic Waste by 2030." Available online: https://www.canada.ca/en/environment-climate-change/news/2020/10/canada-one-step-closer-to-zero-plastic-waste-by-2030.html (accessed on 1 June 2023).

Ernst & Young Global Limited. "Italy Postpones Plastic Packaging Tax to 2022." (2021) Available online: https://www.ey.com/en_gl/taxalerts/italy-postpones-plastic-packaging-tax-to-2022 (accessed on 4 June 2023).

Frias, João PGL, and Roisin Nash. "Microplastics: Finding a consensus on the definition." *Marine Pollution Bulletin* 138 (2019): 145–147.

Galgani, Francois, et al. "Marine litter within the European marine strategy framework directive." *ICES Journal of Marine Science* 70.6 (2013): 1055–1064.

He, Defu, et al. "Microplastics in soils: Analytical methods, pollution characteristics and ecological risks." *TrAC Trends in Analytical Chemistry* 109 (2018): 163–172.

Hurley, Rachel R., et al. "Validation of a method for extracting microplastics from complex, organic-rich, environmental matrices." *Environmental Science & Technology* 52.13 (2018): 7409–7417.

Iwalaye, Oladimeji Ayo, Ganas Kandasamy Moodley, and Deborah Vivienne Robertson-Andersson. "The possible routes of microplastics uptake in sea cucumber Holothuria cinerascens (Brandt, 1835)." *Environmental Pollution* 264 (2020): 114644.

Jewett, Elysia, et al. "Microplastics and their impact on reproduction—Can we learn from the C. elegans model?" *Frontiers in Toxicology* 4 (2022): 748912.

Jiang, Xiaofeng, et al. "Toxicological effects of polystyrene microplastics on earthworm (Eisenia fetida)." *Environmental Pollution* 259 (2020): 113896.

Kershaw, Peter J., and Chelsea M. Rochman. "Sources, fate and effects of microplastics in the marine environment: Part 2 of a global assessment." *Reports and Studies-IMO/FAO/Unesco-IOC/WMO/IAEA/UN/UNEP Joint Group of Experts on the Scientific Aspects of Marine Environmental Protection (GESAMP) Eng No. 93* (2015).

Klöckner, Philipp, Thorsten Reemtsma, and Stephan Wagner. "The diverse metal composition of plastic items and its implications." *Science of the Total Environment* 764 (2021): 142870.

Koelmans, Albert A., et al. "Risks of plastic debris: Unravelling fact, opinion, perception, and belief." *Environmental Science & Technology* 51 (2017): 11513–11519.

Kokalj, Anita Jemec, et al. "Effects of microplastics from disposable medical masks on terrestrial invertebrates." *Journal of Hazardous Materials* 438 (2022): 129440.

Kokalj, Anita Jemec, et al. "Plastic bag and facial cleanser derived microplastic do not affect feeding behaviour and energy reserves of terrestrial isopods." *Science of the Total Environment* 615 (2018): 761–766.

Koyilath Nandakumar, Vaishnavi, Sankar Ganesh Palani, and Murari Raja Varma. "Interactions between microplastics and unit processes of wastewater treatment plants: A critical review." *Water Science and Technology* 85.1 (2022): 496–514.

Kristanti, Risky Ayu, et al. "Overview of microplastics in the environment: Type, source, potential effects and removal strategies." *Bioprocess and Biosystems Engineering* 46.3 (2023): 429–441.

Lahive, Elma, et al. "Microplastic particles reduce reproduction in the terrestrial worm Enchytraeus crypticus in a soil exposure." *Environmental Pollution* 255 (2019): 113174.

Lam, Chung-Sum, et al. "A comprehensive analysis of plastics and microplastic legislation worldwide." *Water, Air, & Soil Pollution* 229 (2018): 1–19.

Lee, Kyun-Woo, et al. "Size-dependent effects of micro polystyrene particles in the marine copepod Tigriopus japonicus." *Environmental Science & Technology* 47.19 (2013): 11278–11283.

Li, Aoyun, et al. "Environmental microplastics exposure decreases antioxidant ability, perturbs gut microbial homeostasis and metabolism in chicken." *Science of the Total Environment* 856 (2023): 159089.

Li, Boqing, et al. "Polyethylene microplastics affect the distribution of gut microbiota and inflammation development in mice." *Chemosphere* 244 (2020): 125492.

Li, Jia, et al. "Effects of microplastics on higher plants: A review." *Bulletin of Environmental Contamination and Toxicology* 109.2 (2022a): 241–265.

Li, Jiana, et al. "Microplastics in mussels along the coastal waters of China." *Environmental Pollution* 214 (2016): 177–184.

Li, Wenlu, et al. "Impacts of microplastics addition on sediment environmental properties, enzymatic activities and bacterial diversity." *Chemosphere* 307 (2022b): 135836.

Liu, Ge, et al. "Microplastic impacts on microalgae growth: Effects of size and humic acid." *Environmental Science & Technology* 54.3 (2019): 1782–1789.

Lo, Hau Kwan Abby, and Kit Yu Karen Chan. "Negative effects of microplastic exposure on growth and development of Crepidula onyx." *Environmental Pollution* 233 (2018): 588–595.

Madigele, Patricia K., Goemeone E. J. Mogomotsi, and Mavis Kolobe. "Consumer willingness to pay for plastic bags levy and willingness to accept eco-friendly alternatives in Botswana." *Chinese Journal of Population Resources and Environment* 15.3 (2017): 255–261.

Mao, Yufeng, et al. "Phytoplankton response to polystyrene microplastics: Perspective from an entire growth period." *Chemosphere* 208 (2018): 59–68.

MESTECC. "Malaysia's Roadmap towards Zero Disposable Plastic Use 2018–2030." (2019). Available online: https://www.pmo.gov.my/ms/2019/07/pelan-hala-tuju-malaysia-ke-arah-sifar-penggunaan-plastik-sekali-guna-2018-2030/ (accessed on 28 May 2023).

Miller, Michaela E., Mark Hamann, and Frederieke J. Kroon. "Bioaccumulation and biomagnification of microplastics in marine organisms: A review and meta-analysis of current data." *PLoS One* 15.10 (2020): e0240792.

Mohsen, Mohamed, et al. "Mechanism underlying the toxicity of the microplastic fibre transfer in the sea cucumber Apostichopus japonicus." *Journal of Hazardous Materials* 416 (2021): 125858.

Mossman, Susan. "Conservation of plastics: Materials science, degradation and preservation." *Nature* 455.7211 (2008): 288–290.

Muposhi, Asphat, Mercy Mpinganjira, and Marius Wait. "Efficacy of plastic shopping bag tax as a governance tool: Lessons for South Africa from Irish and Danish success stories." *Acta Commercii-Independent Research Journal in the Management Sciences* 21.1 (2021): 891.

NCSL. "State Plastic Bag Legislation." (2021). Available online: https://www.ncsl.org/research/environment-and-natural-resources/plastic-bag-legislation.aspx (accessed on 4 June 2023).

Nyathi, Brian, and Chamunorwa Aloius Togo. "Overview of legal and policy framework approaches for plastic bag waste management in African countries." *Journal of Environmental and Public Health* 2020 (2020): 1–8.

O'Brien, Joshua, and Gladman Thondhlana. "Plastic bag use in South Africa: Perceptions, practices and potential intervention strategies." *Waste Management* 84 (2019): 320–328.

Oyake-Ombis, Leah, Bas JM van Vliet, and Arthur PJ Mol. "Managing plastic waste in East Africa: Niche innovations in plastic production and solid waste." *Habitat International* 48 (2015): 188–197.

Pandey, Bhamini, et al. "Microplastics in the ecosystem: An overview on detection, removal, toxicity assessment, and control release." *Water* 15.1 (2022): 51.

Parolini, Marco, et al. "A global perspective on microplastic bioaccumulation in marine organisms." *Ecological Indicators* 149 (2023): 110179.

Plastics Europe Publications. (2020) Plastics Europe Org. Available from: https://www.plasticseurope.org/en/resources/publications/4312-plastics-facts-2020.

Shen, Maocai, et al. "Can microplastics pose a threat to ocean carbon sequestration?" *Marine Pollution Bulletin* 150 (2020): 110712.

Shin, Sun-Kyoung, et al. "New policy framework with plastic waste control plan for effective plastic waste management." *Sustainability* 12.15 (2020): 6049.

Smith, Louise. "Plastic waste." *House of Commons Briefing Paper* 08515 (2020).

"Sweden: Parliament Votes to Adopt Tax on Plastic Bags." (2020) Available online: https://www.loc.gov/item/global-legal-monitor/2020-01-31/sweden-parliament-votes-to-adopt-tax-on-plastic-bags/ (accessed on 4 June 2023).

Thompson, Richard C. "Microplastics in the marine environment: Sources, consequences and solutions." *Marine Anthropogenic Litter* (2015): 185–200. https://doi.org/10.1007/978-3-319-16510-3_7.

Thompson, Richard C., et al. "Lost at sea: Where is all the plastic?" *Science* 304.5672 (2004): 838.

Van Cauwenberghe, Lisbeth, et al. "Microplastic pollution in deep-sea sediments." *Environmental Pollution* 182 (2013): 495–499.

van Wezel, Annemarie, Inez Caris, and Stefan AE Kools. "Release of primary microplastics from consumer products to wastewater in the Netherlands." *Environmental Toxicology and Chemistry* 35.7 (2016): 1627–1631.

Von Moos, Nadia, Patricia Burkhardt-Holm, and Angela Köhler. "Uptake and effects of microplastics on cells and tissue of the blue mussel Mytilus edulis L. after an experimental exposure." *Environmental Science & Technology* 46.20 (2012): 11327–11335.

Worm, Boris, et al. "Plastic as a persistent marine pollutant." *Annual Review of Environment and Resources* 42 (2017): 1–26.

Wright, Stephanie L., Richard C. Thompson, and Tamara S. Galloway. "The physical impacts of microplastics on marine organisms: A review." *Environmental Pollution* 178 (2013): 483–492.

Yu, Qing, et al. "Distribution, abundance and risks of microplastics in the environment." *Chemosphere* 249 (2020): 126059.

Zeng, Eddy Y., ed. *Microplastic Contamination in Aquatic Environments: An Emerging Matter of Environmental Urgency*. Elsevier, Cambridge, 2018.

Zhang, Kai, et al. "Microplastic pollution in China's inland water systems: A review of findings, methods, characteristics, effects, and management." *Science of the Total Environment* 630 (2018): 1641–1653.

Zhang, Qun, et al. "A review of microplastics in table salt, drinking water, and air: Direct human exposure." *Environmental Science & Technology* 54.7 (2020): 3740–3751.

Zimmermann, Lisa, et al. "What are the drivers of microplastic toxicity? Comparing the toxicity of plastic chemicals and particles to Daphnia magna." *Environmental Pollution* 267 (2020): 115392.

4 Bioindicators of Microplastics

Kassian T.T. Amesho, Sumarlin Shangdiar, Chandra Mohan, Chingakham Chinglenthoiba, Ashutosh Pandey, Timoteus Kadhila, Sioni Iikela, and Mohd Nizam Lani

4.1 INTRODUCTION

Microplastic pollution has become a pressing environmental concern worldwide, posing significant threats to both aquatic and terrestrial ecosystems. Microplastics, defined as plastic particles less than 5 mm in size, originate from the fragmentation and weathering of larger plastic debris or are intentionally produced as microbeads for various industrial and consumer applications (Chinglenthoiba et al., 2022). Their ubiquitous presence in natural environments has raised alarm due to potential adverse effects on biota and ecosystems. The impacts of microplastic pollution on various organisms and ecosystems are diverse and multifaceted. Studies have revealed that microplastics can be ingested by a wide range of organisms across various trophic levels, including plankton, fish, birds, and mammals (Liwarska-Bizukoj et al., 2023; Sharma and Garg, 2018). Upon ingestion, these minute plastic particles can lead to a plethora of detrimental consequences. Physical damage to tissues, obstruction of feeding pathways, and potential leaching of toxic chemicals from microplastics into the organisms have been reported (Zhang et al., 2018; Bour et al., 2018). Moreover, microplastics have been implicated as vectors for transporting harmful pollutants and pathogens, potentially magnifying the ecological implications of their presence (Ding et al., 2018).

Given the complex and far-reaching impacts of microplastics on biota and ecosystems, there is an urgent need to develop effective tools for monitoring and assessing microplastic pollution. Bioindicators have emerged as a valuable approach in this endeavor. Bioindicators are organisms or biological responses that can reflect changes in the environment, acting as early warning signals of ecological stress or pollution (Kumar et al., 2018; Christoph et al., 2017). They offer insights into the occurrence and ecological consequences of microplastic contamination, aiding in the understanding of the extent of contamination, the pathways of ingestion, and the potential for trophic transfer within ecosystems. The importance of developing bioindicators lies in their potential to enhance our understanding of the effects of microplastic pollution on various organisms and ecosystems. By studying the responses of different species to microplastic exposure, valuable information about the susceptibility and impacts of microplastics on biota can be gained (Xu et al.,

DOI: 10.1201/9781003438793-4

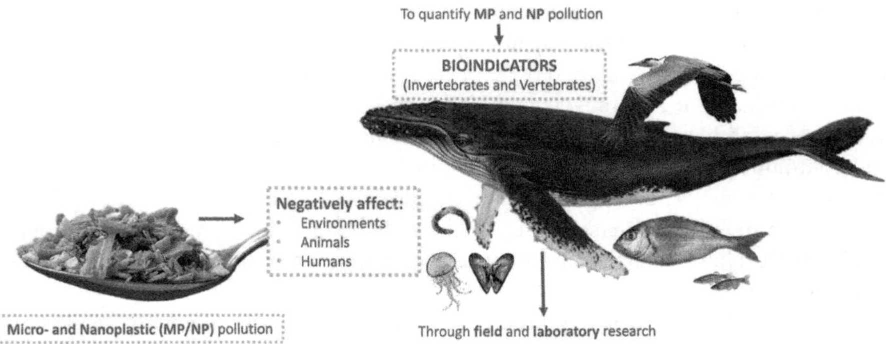

FIGURE 4.1 The role of bioindicators in monitoring of microplastics. (Obtained with permission from Multisanti et al. (2022). Copyright (2022), Elsevier [License Number: 5653810105543].)

2020; Sathish et al., 2020). This knowledge is crucial for identifying vulnerable species and ecosystems that may require special conservation efforts and resource allocation (Jiang et al., 2020).

The scope of this chapter is to provide a comprehensive overview of the different bioindicators that have been identified for monitoring microplastic pollution in aquatic and terrestrial ecosystems. This review will encompass various organisms, including bacteria, algae, invertebrates, and vertebrates, which play crucial roles in different ecological niches and food chains. Moreover, the chapter will explore the methodologies and techniques used to detect and quantify microplastic ingestion in these organisms, such as microscopy, chemical analysis, and biomarkers (Sobhani et al., 2022; Barboza et al., 2018; de Oliveira et al., 2021). Understanding these analytical approaches is vital for accurate and reliable assessments of microplastic contamination in different environments.

Additionally, the chapter will critically evaluate the potential applications and limitations of using bioindicators for monitoring microplastic pollution. This evaluation will pave the way for the integration of these tools into robust environmental monitoring programs. Ultimately, the aim is to contribute to a comprehensive understanding of microplastic pollution and its impacts on biota, facilitating informed decision-making for sustainable environmental management and conservation strategies (Triebskorn et al., 2019). Figure 4.1 shows the conceptual diagram depicting the role of bioindicators in monitoring of microplastics.

4.2 BIOINDICATORS OF MICROPLASTICS IN BACTERIA AND ALGAE

Microplastics' pervasive presence in the environment has raised concerns about their impacts on biota. Among the diverse array of organisms affected, bacteria and algae stand out as promising bioindicators for monitoring microplastic pollution. These microscopic organisms play crucial roles in ecosystem processes and serve as fundamental components of food webs, making them valuable indicators of environmental changes caused by microplastics. Bacteria, as ubiquitous microorganisms, are integral to various ecological functions, including nutrient cycling, organic matter degradation, and biogeochemical processes (Horton et al., 2017). In recent years, research has shed light on the interactions between microplastics and bacterial communities, making bacteria promising candidates for bioindicators of microplastic pollution. When exposed to microplastics, bacteria can colonize and form biofilms on the plastic surface, altering their abundance and composition (Akindele et al., 2019). These changes can serve as indicators of the presence and distribution of microplastics in aquatic environments.

Similarly, algae, as primary producers, play a pivotal role in aquatic ecosystems by fixing carbon through photosynthesis and forming the basis of the food chain (Acquah et al., 2021). Due to their crucial position, algae can act as sentinels, reflecting the influence of microplastics on primary productivity and trophic dynamics. Microplastic particles can interact with algal cells, affecting their growth, physiology, and cellular functions (Abbasi et al., 2018). Therefore, changes in algal responses to microplastic exposure can be utilized as indicators of microplastic contamination in aquatic habitats.

Detecting microplastic ingestion in bacteria and algae poses unique challenges due to their small size and subtle interactions with microplastic particles. Nonetheless, researchers have developed several methodologies to address these challenges. Microscopy techniques, such as light microscopy and electron microscopy, have been widely employed to observe the interactions between microplastics and bacterial cells (Saad et al., 2022). Using fluorescently labeled microplastics, researchers have successfully visualized the adherence and internalization of microplastics by bacterial communities (Zhang et al., 2020). Similarly, algae have been subjected to similar microscopy techniques to identify microplastic ingestion by algal cells (Atamanalp et al., 2022). In addition to microscopy, chemical analysis has emerged as a valuable tool for detecting and quantifying microplastics in bacterial and algal samples. Fourier-transform infrared spectroscopy (FTIR) and Raman spectroscopy are commonly used methods for chemical identification of microplastic particles (Chinglenthoiba et al., 2022). These techniques can be applied to determine the presence of microplastics in bacterial cultures and algal biomass, offering complementary information to microscopy-based approaches.

Several studies have successfully identified microplastic ingestion in bacteria and algae, further emphasizing their potential as bioindicators. For instance, a study by Zhang et al. (2020) revealed significant alterations in bacterial community structure upon exposure to microplastics, providing evidence of microbial responses to microplastic contamination in aquatic systems. In another investigation, Barboza et al. (2020) detected microplastic particles inside algal cells, indicating that algae can uptake microplastics and potentially transfer them to higher trophic levels. Moreover, research has demonstrated that microplastic ingestion can have cascading effects on algal communities. Ríos et al. (2020) found that microplastic exposure led to changes in algal physiology and decreased growth rates, influencing ecosystem dynamics. Overall, the use of bacteria and algae as bioindicators offers valuable insights into the extent and consequences of microplastic pollution in aquatic environments, thereby contributing to our understanding of the ecological implications of microplastic contamination.

4.3 BIOINDICATORS OF MICROPLASTICS IN INVERTEBRATES

The presence of microplastics in aquatic environments has become a growing concern due to their potential impacts on ecosystems and human health. Among the organisms vulnerable to microplastic exposure, invertebrates, particularly molluscs and crustaceans, have drawn significant attention as potential bioindicators of this pollution. This section provides a comprehensive discussion on the various invertebrate species identified as bioindicators of microplastics, the detection methods employed, and examples of studies that have confirmed microplastic ingestion in these organisms. Invertebrates constitute a vital component of aquatic ecosystems, serving as essential links in food chains and playing crucial roles in nutrient cycling. Within this diverse group, mollusks and crustaceans have emerged as promising bioindicators due to their wide distribution, diverse feeding habits, and capacity to filter-feed as indicated in Table 4.1. Additionally, their sedentary nature and longevity in marine and freshwater habitats make them particularly susceptible to accumulating microplastics over time.

Mollusks, including bivalves (e.g., mussels, oysters, clams) and gastropods (e.g., snails), have been extensively studied for their ability to accumulate microplastics. The filter-feeding mechanism of bivalves exposes them to suspended particles, leading to inadvertent ingestion of microplastics alongside their natural diet. Similarly, gastropods, which graze on biofilm and detritus, may inadvertently ingest microplastics adhered to these substrates (Park et al., 2020).

Crustaceans, such as crabs, lobsters, and shrimp, are also potential bioindicators of microplastics. As omnivores and scavengers, they can encounter microplastics directly or indirectly through the consumption of contaminated prey. The ingestion of microplastics may lead to adverse effects on their health and survival, making them valuable indicators of microplastic pollution in aquatic systems (Saley et al., 2019).

4.3.1 METHODS FOR DETECTING MICROPLASTIC INGESTION

Detecting microplastics in invertebrates presents a unique set of challenges, given their small size and the potential for false positives due to microplastic contamination in laboratory settings. To address these issues, various methods have been employed to accurately identify and quantify microplastics in these organisms.

Microscopy: Optical microscopy is a fundamental tool for visualizing and identifying microplastics in invertebrate tissues. Researchers often use staining techniques to enhance the contrast between microplastics and biological components. However, the small size of microplastics, especially in nano-range, may require advanced microscopy techniques, such as scanning electron microscopy (SEM) or Raman spectroscopy, to achieve higher resolution and more precise characterization (Chinglenthoiba et al., 2022).

Biomarkers: Biomarker analysis is another valuable approach used to detect microplastic exposure in invertebrates. Researchers look for specific cellular, molecular, or biochemical changes induced by microplastic ingestion. For instance, oxidative stress biomarkers, such as superoxide dismutase (SOD) and catalase (CAT) activity, may indicate cellular damage caused by microplastics. These biomarkers provide crucial insights into the sublethal effects of microplastic exposure on invertebrates (Gao et al., 2022).

TABLE 4.1
Overview of Various Invertebrates Identified as Potential Bioindicators of Microplastics

Invertebrate Group	Representative Species	Habitat	Feeding Habit	Potential Role as Bioindicator	References
Mollusks	Bivalves (e.g., mussels, oysters, clams)	Estuaries, coastal areas	Filter-feeders on suspended particles	Accumulate microplastics in their tissues due to filter-feeding mechanisms, providing insights into microplastic pollution in marine environments.	Saley et al. (2019)
	Gastropods (e.g., snails)	Freshwater, marine areas	Grazers on biofilm and detritus	May ingest microplastics adhered to their food sources, serving as indicators of microplastic pollution in different habitats.	Saley et al. (2019), Tien et al. (2020)
Crustaceans	Crabs (e.g., blue crabs, shore crabs)	Coastal areas, estuaries	Omnivores and scavengers	Indicate microplastic pollution through ingestion of contaminated prey and direct exposure to microplastics.	Herrera et al. (2019)
	Lobsters (e.g., American lobster)	Marine areas	Opportunistic feeders	Accumulate microplastics in their digestive tracts, reflecting the prevalence of microplastic pollution in marine ecosystems.	Borges-Ramírez et al. (2020), Hastuti et al. (2019)
	Shrimp (e.g., brown shrimp, pink shrimp)	Coastal areas	Filter-feeders on plankton	May ingest microplastics through filter-feeding, indicating the presence of microplastics in their environment.	Alomar et al. (2020)
Other Organisms	Polychaetes (e.g., ragworms)	Sediments, coastal areas	Deposit-feeders and scavengers	Ingest microplastics present in sediments and reflect microplastic pollution in benthic habitats.	Liwarska-Bizukoj et al. (2023), Sobhani et al. (2022), Baalkhuyur et al. (2020)
	Sponges	Marine areas	Filter-feeders on suspended particles	Accumulate microplastics within their tissues, providing insights into microplastic pollution in marine ecosystems.	Chan et al. (2019)
	Sea cucumbers (e.g., Holothuria spp.)	Benthic habitats	Detritivores	Ingest microplastics present in sediment and detritus, acting as indicators of microplastic pollution in marine benthic environments.	Chan et al. (2019), Ding et al. (2020)

Invertebrates, such as mollusks and crustaceans, offer valuable insights into the presence and impact of microplastics in aquatic ecosystems. Their ability to accumulate microplastics, combined with appropriate detection methods and well-designed studies, make them reliable bioindicators (Saley et al., 2019). Understanding the effects of microplastic ingestion in these organisms contributes to our broader understanding of microplastic pollution and its potential consequences for entire ecosystems and human health. Further research in this area will continue to shed light on the extent of microplastic contamination and guide conservation efforts to mitigate its adverse effects.

4.4 BIOINDICATORS OF MICROPLASTICS IN VERTEBRATES

The impact of microplastic pollution on aquatic ecosystems and the organisms that inhabit them has become a major environmental concern. Among the diverse range of organisms affected by this pervasive pollution, vertebrates, including fish and birds, have garnered significant attention as potential bioindicators of microplastics (Kumar et al., 2021). This section presents a comprehensive discussion on vertebrate species identified as bioindicators of microplastics, the methods used to detect microplastic ingestion, and examples of studies confirming microplastic presence in these organisms. Vertebrates, as higher trophic-level organisms, play crucial roles in ecosystem dynamics, and their susceptibility to microplastic ingestion raises concerns about the potential for biomagnification in food chains. Among vertebrates, fish and birds have emerged as prominent bioindicators due to their ecological significance, wide distribution, and frequent interactions with aquatic habitats (Kuśmierek and Popiołek, 2020).

Fish: As inhabitants of freshwater and marine environments, fish encounter microplastics through various routes, including ingestion of contaminated prey, filter-feeding on suspended particles, and exposure to microplastics adhered to aquatic substrates. The occurrence of microplastic ingestion in fish species such as salmon, trout, and cod has been documented, indicating their vulnerability to microplastic exposure (Nematollahi et al., 2021). Numerous studies have provided compelling evidence of microplastic ingestion in various invertebrate species, supporting their potential role as bioindicators. For instance, a study by Barboza et al. (2020) examined the presence of microplastics in wild fish from the North East Atlantic Ocean and found microplastics in substantial quantities. The microplastic particles were shown to accumulate in the gut, potentially affecting nutrient absorption and overall health. Similarly, research by Herrera et al. (2019) investigated the presence of microplastics in the digestive tracts of chub mackerel (*Scomber colias*) in the Canary Islands coast. The study revealed a higher incidence of microplastics in chub mackerel come from more urbanized areas, highlighting the impact of human activities on microplastic pollution in marine environments.

Birds: Coastal and marine birds, being top predators in their ecosystems, are also susceptible to microplastic ingestion. They may mistakenly consume floating microplastics, often resembling prey items, while foraging at sea. Additionally, plastic debris, including microplastics, may be regurgitated to nestlings, potentially leading to deleterious effects on their health and survival (Renzi et al., 2019).

4.4.1 Methods for Detecting Microplastic Ingestion

Accurate detection of microplastics in vertebrates requires sophisticated methods to avoid false positives and quantify microplastic load effectively, as indicated in Table 4.2. Researchers have employed various approaches to assess microplastic ingestion in these organisms.

Microscopy: Optical microscopy remains one of the primary methods for visually identifying microplastics in vertebrate tissues. Depending on the size and type of microplastics, researchers may use light microscopy or advanced techniques such as Fourier-transform infrared spectroscopy (FTIR) to differentiate microplastics from organic matter (Chinglenthoiba et al., 2022).

Chemical Analysis: Chemical techniques, such as Raman spectroscopy and gas chromatography-mass spectrometry (GC-MS), are utilized to confirm the presence of microplastics and determine their polymer composition (Chinglenthoiba et al., 2022). This approach provides quantitative data on microplastic types and concentrations, aiding in understanding the extent of ingestion and potential impacts on vertebrate health.

Numerous studies have demonstrated microplastic ingestion in various vertebrate species, reinforcing their role as bioindicators of microplastic pollution. A study conducted by Borges-Ramírez et al. (2020) investigated the occurrence of microplastics in the gastrointestinal tracts of several fish species from different coastal areas. The results indicated widespread ingestion of microplastics across the sampled populations, highlighting the pervasive nature of this pollution in marine ecosystems. In another study by Patterson et al. (2019), the stomach contents of marine birds from different regions were analyzed, revealing microplastic ingestion in several avian species. The study emphasized the potential risks of microplastic exposure to bird populations and underscored the importance of mitigating this environmental threat. Furthermore, a study by Merga et al. (2020) examined microplastic ingestion in freshwater fish species inhabiting contaminated rivers. The research elucidated the impact of urbanization and industrial activities on microplastic prevalence in aquatic environments and its potential implications for fish health and the broader ecosystem.

Vertebrates, particularly fish and birds, serve as valuable bioindicators of microplastic pollution, reflecting the extent of contamination in aquatic ecosystems. Employing reliable detection methods is crucial to accurately quantify microplastic ingestion and understand its ecological implications. The presence of microplastics in these higher trophic-level organisms raises concerns about the potential for biomagnification, with potential repercussions on human health through the consumption of contaminated seafood. Robust research in this field is vital to inform sustainable policies and actions aimed at reducing microplastic pollution and safeguarding the health of aquatic ecosystems and the species that depend on them (Ghosh et al., 2021).

TABLE 4.2
Methods for Detecting Microplastic Ingestion in Organisms

Detection Method	Description	Advantages	Disadvantages	References
Microscopy	Optical microscopy is a traditional technique used to visualize and identify microplastics in organisms. Sample preparation involves dissecting or homogenizing tissues, followed by visual inspection using a microscope. Staining techniques may be employed to enhance contrast between microplastics and organic matter.	Provides direct visual evidence of microplastics.	Limited ability to distinguish microplastics from other small particles or fibers.	Horton et al. (2017)
Raman Spectroscopy	Raman spectroscopy is a chemical technique used to analyze the molecular structure of materials. It can identify microplastics based on their unique Raman spectra, allowing differentiation from other particles.	Non-destructive analysis, preserving sample integrity.	Requires expensive equipment and skilled operators.	Chinglenthoiba et al. (2022)
Fourier-Transform Infrared Spectroscopy (FTIR)	FTIR is another chemical technique that identifies microplastics by measuring the absorption of infrared light by specific chemical bonds in their polymers. It allows identification of different polymer types in a sample.	Accurate identification of microplastic polymers.	Cannot determine the size and shape of microplastics.	Chinglenthoiba et al. (2022)
Gas Chromatography-Mass Spectrometry (GC-MS)	GC-MS is a powerful analytical technique used to separate and identify complex mixtures of organic compounds. It is utilized to detect and quantify microplastic monomers, additives, and breakdown products in organisms.	Provides detailed information on microplastic chemical composition.	Requires extensive sample preparation and specialized equipment.	Chinglenthoiba et al. (2022)
Fluorescence Microscopy	Fluorescence microscopy utilizes the emission of fluorescent light by microplastics when exposed to specific wavelengths of light. It is valuable for detecting small microplastics or particles in complex matrices.	Enhanced sensitivity to detect small microplastics.	Requires specific fluorescent labeling or tagging of microplastics.	Patterson et al. (2020), Miller et al. (2020)
Spectrofluorometry	Spectrofluorometry is a technique used to measure the fluorescence emission of samples when excited by specific wavelengths of light. It is employed to quantify fluorescently tagged microplastics in organisms.	Enables quantitative analysis of fluorescently tagged microplastics.	Limited to detecting only fluorescently tagged microplastics.	Karami et al. (2017), Chinglenthoiba et al. (2022)
Chemical Digestion	Chemical digestion involves treating samples with strong acids or enzymes to dissolve organic matter, leaving microplastics unaffected. This method allows separation and quantification of microplastics from biological tissues.	Effective separation of microplastics from biological tissues.	May lead to sample contamination and alteration of microplastic properties.	Patterson et al. (2019), Merga et al. (2020)

4.5 POTENTIAL APPLICATIONS OF BIOINDICATORS FOR MONITORING MICROPLASTIC POLLUTION

As microplastic pollution continues to pose a significant threat to aquatic ecosystems and human health, the utilization of bioindicators presents a promising avenue for monitoring and understanding its distribution, accumulation, and impacts (Koongolla et al., 2020). This section discusses the potential applications of bioindicators in monitoring microplastic pollution, focusing on their role in identifying hotspots and sources of pollution. Additionally, the limitations and challenges associated with employing bioindicators for monitoring microplastic pollution are explored to gain a comprehensive understanding of their applicability.

Bioindicators offer valuable insights into the spatial distribution of microplastics in ecosystems, aiding in the identification of pollution hotspots. By assessing the abundance and diversity of microplastic-ingesting organisms in different regions, researchers can pinpoint areas where microplastic contamination is particularly prevalent. For instance, the high abundance of microplastic-ingesting fish and birds in specific coastal areas may indicate hotspots of pollution (Tien et al., 2020). Integrating data from multiple bioindicators can help establish a more comprehensive understanding of the extent and variability of microplastic pollution within an ecosystem. Bioindicators can provide crucial information on the sources of microplastic pollution in aquatic environments. Since different types of microplastics are associated with various sources (e.g., microbeads in personal care products, fibers from textiles, fragments from plastic debris), analyzing the composition and characteristics of microplastics ingested by bioindicators can offer insights into potential pollution sources. Additionally, stable isotope analysis and chemical fingerprinting of microplastics in bioindicators may help trace the origin of microplastics and their pathways through food webs (Borges-Ramírez et al., 2020). Understanding the sources of microplastic pollution is essential for implementing targeted mitigation strategies and policies.

4.6 LIMITATIONS AND CHALLENGES

One of the significant challenges in using bioindicators for monitoring microplastic pollution is the species-specific response to microplastic exposure. Different organisms may exhibit varying levels of ingestion, retention, and adverse effects of microplastics. Moreover, the sensitivity of bioindicators to microplastic ingestion can be influenced by factors such as life stage, feeding habits, and habitat preferences (Multisanti et al., 2022). Therefore, extrapolating results from one bioindicator to represent an entire ecosystem requires careful consideration and validation through comprehensive studies encompassing various species. Monitoring microplastic pollution using bioindicators is further complicated by temporal and spatial variability. The abundance of microplastics and bioindicator species may fluctuate seasonally and vary across different regions due to changing environmental conditions and human activities. Thus, long-term monitoring efforts and large-scale studies are essential to account for these variations and establish reliable trends in microplastic pollution.

Aquatic ecosystems are subjected to a myriad of stressors, including chemical pollutants, climate change, and habitat destruction, which can interact with microplastic pollution and influence the responses of bioindicators. Understanding the specific impacts of microplastics in the context of other stressors is challenging, and separating their individual effects requires careful experimental design and data analysis. Moreover, confounding factors, such as the ingestion of natural particles resembling microplastics, can lead to false positive results and require rigorous validation of detection methods (Borges-Ramírez et al., 2020).

Bioindicators offer valuable applications in monitoring microplastic pollution, providing essential information on pollution hotspots and potential sources of contamination in aquatic ecosystems (Patterson et al., 2019). However, their use requires careful consideration of species-specific responses, temporal and spatial variability, and the influence of multiple stressors. As an integrated tool in environmental monitoring, bioindicators complement physical and chemical analyses, contributing to a comprehensive understanding of microplastic pollution and informing strategies for mitigating its impacts on ecosystems and society.

4.7 CONCLUSIONS AND WAY FORWARD

Microplastic pollution has emerged as a pressing environmental concern, posing substantial risks to aquatic ecosystems and human well-being. In this chapter, we explored the role of bioindicators in understanding and monitoring microplastic pollution, with a focus on their identification, detection methods, and potential applications. Through extensive research and the collective efforts of the scientific community, the significance of bioindicators in elucidating the presence and impacts of microplastic pollution has been unequivocally demonstrated. Detection methods such as microscopy, spectroscopy, and chemical analysis have been instrumental in accurately quantifying microplastic ingestion in various organisms. However, challenges associated with distinguishing microplastics from natural particles, species-specific responses to microplastics, and temporal and spatial variability in pollution call for further advancements in detection technologies. Bioindicators play pivotal roles in identifying hotspots of contamination and providing insights into the sources of microplastic pollution. Integration of bioindicator data with physical and chemical analyses enhances the comprehensiveness of environmental monitoring programs, leading to a more holistic understanding of microplastic pollution in ecosystems. Despite advancements in the field, several crucial avenues of research demand attention. Standardization of methods is urgently needed to ensure comparability of data across studies and the establishment of global databases for microplastic pollution. Biomarkers hold immense potential as indicators of microplastic exposure and its effects on organisms, and necessitating research focused on identifying robust and specific biomarkers. Long-term monitoring studies are essential to comprehensively assess the temporal variability of microplastic pollution and its long-term effects on ecosystems, providing crucial data for identifying trends and predicting future scenarios.

In light of the overwhelming evidence of the pervasiveness and impact of microplastic pollution, concerted action is imperative. Collaboration among researchers, institutions, and governments should be fostered to pool resources and expertise on microplastic pollution. Public awareness about microplastic pollution and its consequences should be increased, empowering individuals to make informed choices in reducing plastic consumption. Advocacy for stricter regulations and incentives for sustainable practices in plastic production, use, and waste management, along with the encouragement of innovation in biodegradable materials and eco-friendly alternatives, will pave the way toward a plastic-free future. In conclusion, bioindicators are indispensable tools in understanding and monitoring microplastic pollution, and by harnessing their potential and addressing research gaps, we can collectively safeguard our ecosystems, protect biodiversity, and secure a healthier environment for future generations.

REFERENCES

Abbasi, S., Soltani, N., Keshavarzi, B., Moore, F., Turner, A., & Hassanaghaei, M., (2018). Microplastics in different tissues of fish and prawn from the Musa Estuary, Persian Gulf. *Chemosphere* 205, 80–87. https://doi.org/10.1016/J.CHEMOSPHERE.2018.04.076.

Acquah, J., Liu, H., Hao, S., Ling, Y., & Ji, J., (2021). Microplastics in freshwater environments and implications for aquatic ecosystems: A mini review and future directions in Ghana. *J. Geosci. Environ. Prot.* 9, 58–74. https://doi.org/10.4236/gep.2021.93005.

Akindele, E. O., Ehlers, S. M., & Koop, J. H. E. (2019). First empirical study of freshwater microplastics in West Africa using gastropods from Nigeria as bioindicators. *Limnologica* 78. https://doi.org/10.1016/j.limno.2019.125708.

Alomar, C., Deudero, S., Compa, M., & Guijarro, B. (2020). Exploring the relation between plastic ingestion in species and its presence in seafloor bottoms. *Mar. Pollut. Bull.* 160, 111641.

Atamanalp, M., Koktürk, M., Parlak, V., Ucar, A., Arslan, G., & Alak, G. (2022). A new record for the presence of microplastics in dominant fish species of the Karasu River Erzurum, Turkey. *Environ. Sci. Pollut. Res.* 29, 7866–7876. https://doi.org/10.1007/s11356-021-16243-w.

Baalkhuyur, F. M., Qurban, M. A., Panickan, P., & Duarte, C. M. (2020). Microplastics in fishes of commercial and ecological importance from the Western Arabian Gulf. *Mar. Pollut. Bull.* 152, 110920.

Barboza, L. G. A., Lopes, C., Oliveira, P., Bessa, F., Otero, V., Henriques, B., Raimundo, J., Caetano, M., Vale, C., & Guilhermino, L. (2020). Microplastics in wild fish from North East Atlantic Ocean and its potential for causing neurotoxic effects, lipid oxidative damage, and human health risks associated with ingestion exposure. *Sci. Total Environ.* 717, 134625. https://doi.org/10.1016/J.SCITOTENV.2019.134625.

Barboza, L. G. A., Vieira, L. R., Branco, V., Figueiredo, N., Carvalho, F., Carvalho, C., & Guilhermino, L. (2018). Microplastics cause neurotoxicity, oxidative damage and energy-related changes and interact with the bioaccumulation of mercury in the European seabass, Dicentrarchus labrax (Linnaeus, 1758). *Aquat. Toxicol.* 195, 49–57. https://doi.org/10.1016/J.AQUATOX.2017.12.008.

Borges-Ramírez, M. M., Mendoza-Franco, E. F., Escalona-Segura, G., & Rendón-von Osten, J. (2020). Plastic density as a key factor in the presence of microplastic in the gastrointestinal tract of commercial fishes from Campeche Bay, Mexico. *Environ. Pollut.* 267, 115659.

Bour, A., Avio, C. G., Gorbi, S., Regoli, F., & Hylland, K. (2018). Presence of microplastics in benthic and epibenthic organisms: Influence of habitat, feeding mode and trophic level. *Environ. Pollut.* 243, 1217–1225. https://doi.org/10.1016/j.envpol.2018.09.115.

Chan, H. S. H., Dingle, C., & Not, C. (2019). Evidence for non-selective ingestion of microplastic in demersal fish. *Mar. Pollut. Bull.* 149, 110523.

Chinglenthoiba, C., Amesho, K. T. T., Reddy, D. G. C. V., et al. (2022). Microplastic as an emerging environmental threat: A critical review on sampling and identification techniques focusing on aquatic ecoystem. *J. Polym. Environ.* 31, 1725–1747. https://doi.org/10.1007/s10924-022-02716-7.

Christoph, D. R., Jahnke, A., Gorokhova, E., Kühnel, D., & Mechthild, S. J. (2017). Impacts of biofilm formation on the fate and potential effects of microplastic in the aquatic environment. *Environ. Sci. Technol. Lett.* 4(7), 258e267. https://doi.org/10.1021/acs.estlett.7b00164.

de Oliveira, J. P. J., Estrela, F. N., Rodrigues, A. S. L., Guimarães, A. T. B., Rocha, T. L., & Malafaia, G. (2021). Behavioral and biochemical consequences of Danio rerio larvae exposure to polylactic acid bioplastic. *J. Hazard. Mater.* 404 (Pt A), 124152. https://doi.org/10.1016/j.jhazmat.2020.124152.

Ding, J., Li, J., Sun, C., Jiang, F., He, C., Zhang, M., Ju, P., & Ding, N. X. (2020). An examination of the occurrence and potential risks of microplastics across various shellfish. *Sci. Total Environ.* 739, 139887.

Ding, J., Zhang, S., Razanajatovo, R. M., Zou, H., & Zhu, W. (2018). Accumulation, tissue distribution, and biochemical effects of polystyrene microplastics in the freshwater fish red tilapia (Oreochromis niloticus). *Environ. Pollut.* 238, 1–9. https://doi.org/10.1016/j.envpol.2018.03.001.

Gao, S., Li, Z., Wang, N., Lu, Y., & Zhang, S. (2022). Microplastics in different tissues of caught fish in the artificial reef area and adjacent waters of Haizhou Bay. *Mar. Pollut. Bull.* 174, 113112.

Ghosh, G. C., Akter, S. M., Islam, R. M., Habib, A., Chakraborty, T. K., Zaman, S., Kabir, A. E., Shipin, O. V., & Wahid, M. A. (2021). Microplastics contamination in commercial marine fish from the Bay of Bengal. *Reg. Stud Mar. Sci.* 44, 101728.

Hastuti, A. R., Lumbanbatu, D. T., & Wardiatno, Y. (2019). The presence of microplastics in the digestive tract of commercial fishes off Pantai Indah Kapuk coast, Jakarta. Indonesia. *Biodivers. J. Biol. Diversity* 20(5), 1232–1242.

Herrera, A., Ŝtindlová, A., Martínez, I., Rapp, J., Romero-Kutzner, V., Samper, M. D., Montoto, T., Aguiar-González, B., Packard, T., & Gómez, M. (2019). Microplastic ingestion by Atlantic chub mackerel (Scomber colias) in the Canary Islands coast. *Mar. Pollut. Bull.* 139, 127–135. https://doi.org/10.1016/J.MARPOLBUL.2018.12.022.

Horton, A. A., Walton, A., Spurgeon, D. J., Lahive, E., & Svendsen, C. (2017). Microplastics in freshwater and terrestrial environments: Evaluating the current understanding to identify the knowledge gaps and future research priorities. *Sci. Total Environ.* 586, 127–141.

Jiang, X., Chang, Y., Zhang, T., Qiao, Y., Klobučar, G., & Li, M. (2020). Toxicological effects of polystyrene microplastics on earthworm (Eisenia fetida). *Environ. Pollut.* 259, 113896. https://doi.org/10.1016/j.envpol.2019.113896.

Karami, A., Golieskardi, A., Choo, C. K., Romano, N., Ho, Y. Bin, & Salamatinia, B. (2017). A high performance protocol for extraction of microplastics in fish. *Sci. Total Environ.* 578, 485–494. https://doi.org/10.1016/j.scitotenv.2016.10.213.

Koongolla, J. B., Lin, L., Pan, Y. F., Yang, C. P., Sun, D. R., Liu, S., Xu, X. R., Maharana, D., Huang, J. S., & Li, H. X. (2020). Occurrence of microplastics in gastrointestinal tracts and gills of fish from Beibu Gulf, South China Sea. *Environ. Pollut.* 258, 113734.

Kumar, R., Sharma, P., Manna, C., & Jain, M. (2021). Abundance, interaction, ingestion, ecological concerns, and mitigation policies of microplastic pollution in riverine ecosystem: A review. *Sci. Total Environ.* 782, 146695. https://doi.org/10.1016/J.SCITOTENV.2021.146695.

Kumar, V. E., Ravikumar, G., & Jeyasanta, K. I. (2018). Occurrence of microplastics in fishes from two landing sites in Tuticorin, South east coast of India. *Mar. Pollut. Bull.* 135, 889–894. https://doi.org/10.1016/j.marpolbul.2018.08.023.

Kuśmierek, N., & Popiołek, M. (2020). Microplastics in freshwater fish from Central European lowland river (Widawa R., SW Poland). *Environ. Sci. Pollut. Res.* 27 (10), 11438–11442. https://doi.org/10.1007/s11356-020-08031-9.

Liwarska-Bizukoj, E., Przemysław Bernat, P., & Jasińska, A. (2023). Effect of bio-based microplastics on earthworms Eisenia Andrei. *Sci. Total Environ.* 898, 165423. http://dx.doi.org/10.1016/j.scitotenv.2023.165423.

Merga, L. B., Redondo-Hasselerharm, P. E., Van den Brink, P. J., & Koelmans, A. A. (2020). Distribution of microplastic and small macroplastic particles across four fish species and sediment in an African lake. *Sci. Total Environ.* 741, 140527.

Miller, M. E., Hamann, M., & Kroon, F. J. (2020). Bioaccumulation and biomagnification of microplastics in marine organisms: A review and meta-analysis of current data. *PLoS ONE* 15(10), e0240792.

Multisanti, C. R., Merola, C., Perugini, M., Aliko, V., Faggio, C. (2022). Sentinel species selection for monitoring microplastic pollution: A review on one health approach. *Ecol. Indic.* 145, 109587. https://doi.org/10.1016/j.ecolind.2022.109587.

Nematollahi, M. J., Keshavarzi, B., Moore, F., Esmaeili, H. R., Nasrollahzadeh Saravi, H., & Sorooshian, A. (2021). Microplastic fibers in the gut of highly consumed fish species from the southern Caspian Sea. *Mar. Pollut. Bull.* 168, 112461. https://doi.org/10.1016/J.MARPOLBUL.2021.112461.

Park, T. J., Lee, S. H., Lee, M. S., Lee, J. K., Lee, S. H., & Zoh, K. D. (2020). Occurrence of microplastics in the Han River and riverine fish in South Korea. *Sci. Total Environ.* 708, 134535. https://doi.org/10.1016/j.scitotenv.2019.134535.

Patterson, J., Jeyasanta, K. I., Sathish, N., Booth, A. M., & Edward, J. P. (2019). Profiling microplastics in the Indian edible oyster, Magallana bilineata collected from the Tuticorin coast, Gulf of Mannar, Southeastern India. *Sci. Total Environ.* 691, 727–735.

Patterson, J., Jeyasanta, K. I., Sathish, N., Edward, J. P., & Booth, A. M. (2020). Microplastic and heavy metal distributions in an Indian coral reef ecosystem. *Sci. Total Environ.* 744, 140706.

Renzi, M., Specchiulli, A., Blašković, A., Manzo, C., Mancinelli, G., & Cilenti, L. (2019). Marine litter in stomach content of small pelagic fishes from the Adriatic Sea: Sardines (Sardina pilchardus) and anchovies (Engraulis encrasicolus). *Environ. Sci. Pollut. Res.* 26 (3), 2771–2781. https://doi.org/10.1007/s11356-018-3762-8.

Ríos, M. F., Hernández-Moresino, R. D., & Galván, D. E. (2020). Assessing urban microplastic pollution in a benthic habitat of Patagonia Argentina. *Mar. Pollut. Bull.* 159. https://doi.org/10.1016/j.marpolbul.2020.111491.

Saad, D., Chauke, P., Cukrowska, E., Richards, H., Nikiema, J., Chimuka, L., Tutu, H. (2022). First biomonitoring of microplastic pollution in the Vaal river using Carp fish (Cyprinus carpio) "as a bio-indicator". *Sci. Total Environ.* 836 (2022) 155623. https://doi.org/10.1016/j.scitotenv.2022.155623.

Saley, A. M., Smart, A. C., Bezerra, M. F., Burnham, T. L. U., Capece, L. R., Lima, L. F. O., Carsh, A. C., Williams, S. L., & Morgan, S. G. (2019). Microplastic accumulation and biomagnification in a coastal marine reserve situated in a sparsely populated area. *Mar. Pollut. Bull.* 146, 54–59.

Sathish, M. N., Jeyasanta, K. I., & Patterson, J. (2020). Monitoring of microplastics in the clam Donax cuneatus and its habitat in Tuticorin coast of Gulf of Mannar (GoM), India. *Environ. Pollut.* 266(1), 115219. https://doi.org/10.1016/j.envpol.2020.115219.

Sharma, K., & Garg, V. K. (2018). Solid-state fermentation for vermicomposting: A step toward sustainable and healthy soil, chapter 17. In A. Pandey, C. Larroche, & C. R. Soccol (Eds.), *Current Developments in Biotechnology and Bioengineering*. Elsevier, pp. 373–413. https://doi.org/10.1016/B978-0-444-63990-5.00017-7.

Sobhani, Z., Panneerselvan, L., Fang, Ch., Naidu, R., & Megharaj, M. (2022). Chronic and transgenerational effects of polyethylene microplastics at environmentally relevant concentrations in earthworms. *Environ. Technol. Innov.* 25, 102226. https://doi.org/10.1016/j.eti.2021.102226.

Tien, C. J., Wang, Z. X., & Chen, C. S. (2020). Microplastics in water, sediment and fish from the Fengshan River system: Relationship to aquatic factors and accumulation of polycyclic aromatic hydrocarbons by fish. *Environ. Pollut.* 265, 114962.

Triebskorn, R., Braunbeck, T., Grummt, T., Hanslik, L., Huppertsberg, S., et al. (2019). Relevance of nano- and microplastics for freshwater ecosystems: A critical review. *TrAC Trends Anal. Chem.* 110, 375–392. https://doi.org/10.1016/J.TRAC.2018.11.023.

Xu, X., Wong, C. Y., Nora, Tam, F. Y., Lo, H. S., & Cheung, S. G. (2020). Microplastics in invertebrates on soft shores in Hong Kong: Influence of habitat, taxa and feeding mode. *Sci. Total Environ.* 715, 136999. https://doi.org/10.1016/j.scitotenv.2020.136999.

Zhang, D., Cui, Y., Zhou, H., Jin, C., Yu, X., Xu, Y., Li, Y., & Zhang, C. (2020). Microplastic pollution in water, sediment, and fish from artificial reefs around the Ma'an Archipelago, Shengsi, China. *Sci. Total Environ.* 703, 134768.

Zhang, S., Yang, X., Gertsen, H., Peters, P., Salanki, T., & Geissen, V. (2018). A simple method for the extraction and identification of light density microplastics from soil. *Sci. Total Environ.* 616, 1056–1065. https://doi.org/10.1016/j.scitotenv.2017.10.213.

5 Role of Various Microbes and Their Enzymatic Mechanisms for Biodegradation of Microplastics

Anik Majumdar

5.1 INTRODUCTION

With an annual production of over 359 million tonnes of synthetic plastics, these materials have become an integral part of our daily lives (Plastics Europe 2019). Assumptions suggest that by 2050, a staggering 12,000 million tonnes of plastic waste will be generated (Geyer et al. 2017). An African scientist originally used the word "microplastic" in his 1990 research, "Plastic and Other Artifacts on South African Beaches: Temporal Trends in Abundance and Composition." Since then, this term has been widely adopted to describe minute plastic particles found across the globe (Alimi et al. 2021). While there is ongoing debate about the specific characteristics of microplastics, most scientists agree that they are defined as plastic particles ranging in size from 100 to 5 mm. The larger plastic particles exceeding 25 mm are referred to as macroplastics, while smaller ones below 100 nm are termed nanoplastics (Yang et al. 2021). The global plastic production has witnessed a rapid increase, surging from 254 million tonnes to 359 million tonnes in just a decade between 2008 and 2018. Predictions indicate that this output will triple by 2050 (Chia et al. 2020). Despite significant efforts in recycling, reuse, and incineration, approximately 32% of all plastic waste still finds its way into the natural environment, particularly terrestrial and aquatic ecosystems (de Souza Machado et al. 2018). Notably, the majority of marine plastic debris, accounting for around 80%, originates from the terrestrial biosphere (Jambeck et al. 2015). Agricultural plastics, known as agroplastics, contribute significantly to plastic pollution in terrestrial ecosystems. Multiple pathways, including the use of plastic coatings for fertilizers (Heuchan et al. 2019), wastewater irrigation (Zhang and Liu 2018), compost applications, biosolids usage (Weithmann et al. 2018), and notably, the utilization of mulching films (Qi et al. 2018) introduce microplastics into agroecosystems. Consequently, the accumulation of plastic in terrestrial ecosystems serves as a long-term reservoir with potential repercussions for freshwater and marine environments. In response to growing concerns about microplastic

DOI: 10.1201/9781003438793-5

contamination in agroecosystems, biodegradable polymers have been developed as an alternative to petroleum-based plastics. Unlike petroleum-based microplastics, biodegradable counterparts can be rapidly degraded by various microbes without releasing harmful by-products (Sander 2019). Therefore, this chapter primarily focuses on the microbial-mediated degradation of microplastics, which holds great promise in the near future.

5.2 MICROPLASTIC FORMATION ROUTE

The two main types of microplastics that are present in the environment are primary and secondary, and their differences can be attributed to where they come from. Primary microplastics are produced during the manufacturing process. For instance, millimeter-sized plastic particles are generated for use in consumer products like toothpaste, facial cleansers, and shower gels. Additionally, primary microplastics can inadvertently result from material abrasion in the production of resin pellets within the air-blasting industry (Nava and Leoni 2021; Khalid et al. 2021). In contrast, secondary microplastics are formed as a consequence of various chemical, physical, and biological processes that break down larger polymer materials into tiny particles. These processes include exposure to ultraviolet (UV) radiation, the freeze–thaw cycle, water turbulence, and the impact of waves (Khalid et al. 2021; Dong et al. 2021).

Microplastic pollution in the environment is predominantly driven by human anthropogenic activities. Given the widespread daily use of plastics, a multitude of industrial and domestic actions contribute to this environmental concern. Notable industries implicated in the presence of microplastics in the environment include raw plastic production, textile manufacturing, and laundry services (Cai et al. 2020; Tang et al. 2021). Furthermore, industrial operations and wastewater treatment facilities (WWTPs) have been identified in numerous studies as major sources of environmental microplastic pollution (Sol et al. 2020). The agricultural sector is also a significant contributor to microplastic contamination (Zurier and Goddard 2021; Kumar et al. 2020). Leachate originating from runoff surface water and improper disposal practices, such as littering, further exacerbates the issue of microplastic pollution (He et al. 2019).

5.3 HARMFUL EFFECTS OF MICROPLASTICS ON HUMAN, SOIL AND PLANT HEALTH

Microplastics (MPs) have been associated with adverse effects on human health, soil quality, and plants. Cox et al. (2019) found that the average person consumes between 39,000 and 52,000 microplastics annually, which can lead to intestinal blockages, inflammatory responses, and disruptions in gut microbiome composition and metabolism. Smaller, less dense particles tend to accumulate in the lungs, where they trigger chronic inflammation through the release of chemotactic molecules (Oliveira et al. 2020). Recent discoveries have revealed the presence of microplastics in human blood and lung tissues (Leslie et al. 2022). Due to their persistence in the body and the release of toxic additives, microplastics can induce oxidative stress and

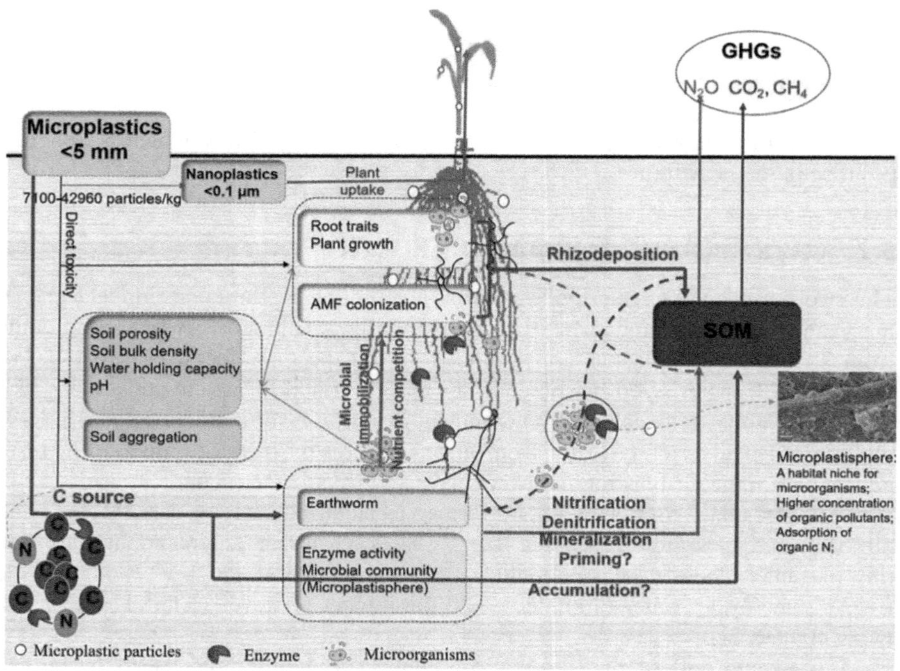

FIGURE 5.1 Effects of microplastics on plant-soil health and soil organic matter decomposition, along with interactions with soil microbes and plants. (Figure adapted from Zhou, J., Wen, Y., Marshall, M. R., Zhao, J., Gui, H., Yang, Y. & Zang, H. (2021b). Microplastics as an emerging threat to plant and soil health in agroecosystems. *Science of the Total Environment*, *787*, 147444 with permission from Elsevier.)

cytotoxicity upon contact with mucous membranes or absorption, leading to inflammation, immune responses, nervous system damage, disrupted metabolism, DNA damage, and potentially cancer (Vethaak and Legler 2021; Gruber et al. 2023).

In soil ecosystems, microplastics can directly or indirectly impact ecosystem processes, plant health, and soil characteristics (Figure 5.1). Their unique properties can alter the physical structure of soil. Microplastics tend to bind with soil minerals and organic components, influencing soil aggregation (Lei et al. 2018). For instance, polylactic acid may increase soil pH, while polyethylene has been observed to reduce it (Boots et al. 2019). Changes in soil pH can have significant implications for the soil microbiome. Microplastics can also affect soil water dynamics. Polyethylene films (2 mm) have been associated with increased soil water loss through elevated evaporation, whereas polyester fibers (8 mm) enhance water-holding capacity, potentially extending soil moisture retention (Wan et al. 2019). Microplastics may serve as a novel ecological niche for microorganisms residing at the soil-plastic interface, creating distinctive microbial communities known as the "microplastisphere" (Zhou et al. 2021a). Studies by Cao et al. (2017) showed that high concentrations of MPs (1%–2%) led to increased mortality among earthworms and hindered their development. Earthworms, by transporting MPs deeper into the soil through their burrows, casts, and surface adhesion, act as vectors that increase the risk of exposure for other soil organisms. Ingestion is a primary factor determining the toxicity of MPs to soil fauna and their potential for bioaccumulation in the food chain (Zhu et al. 2018). Furthermore, MPs can concentrate and retain harmful substances on their surfaces, posing risks to living organisms and humans. For instance, Jeong et al. (2016) found that smaller polystyrene particles (0.05 mm) had a greater impact on rotifer species than larger MPs (0.5 and 6 mm). While there has been substantial research on the impact of MPs on soil fauna, much of it has focused on earthworms. Additional research is needed to explore the interactions between MPs, soil fauna, and other soil microorganisms.

Because of the large molecular weight and size of microplastics, which hinder their ability to pass through the cellulose cell walls of plants, it is commonly believed that plants are unable to absorb them. Nevertheless, MPs have the ability to cross cell membranes and penetrate plant cells when they are reduced to nanoparticles (0.1 nm) and so enter the food chain (Su et al. 2019). In cell culture, Bandmann et al. (2012) showed that tobacco BY-2 cells could absorb nano-polystyrene (0.02 and 0.04 nm). The roots of plants are the most affected by MPs, followed by leaves, shoots, and, finally, stems. This pattern arises from the ease with which plant roots can absorb MPs from contaminated soils and the deposition of MPs from the air onto above-ground plant parts (Zhang et al. 2020). Jiang et al. (2019) found that MPs could impede the above- and below-ground growth of wheat during both vegetative and reproductive phases, leading to delayed germination and toxicity in *Vicia faba*. Phytotoxicity may be attributed to the presence of chemicals like plasticizers and flame retardants integrated into plastics or other secondary pollutants such as heavy metals and antibiotics adsorbed on the surface of MPs (Shen et al. 2019). These substances may be loosely attached or not attached at all to the polymer molecules, allowing them to leach into the soil and adversely affect plant development. Additionally, the size of MPs plays a role in their harmful effects on plants, with smaller MPs causing more damage than larger ones, as reported by Li et al. (2020). In summary, MPs can induce phytotoxicity and directly hinder plant growth by serving as reservoirs and carriers of harmful contaminants within the environment.

5.4 FACTORS AFFECTING MICROBIAL DEGRADATION OF MICROPLASTICS

Microbial degradation of microplastics (MPs) is influenced by a multitude of factors, broadly categorized into two main groups: those affecting the growth and development of microorganisms and those impacting MP properties and the surrounding environment. The chemical and physical characteristics of MPs, including their molecular weight, density, crystallinity, and the presence of functional groups and substituents in their structure, play a pivotal role in microbial degradation. The polymer's morphology, encompassing factors like branching, crystallinity, and physical shape, significantly affects the rate at which plastic polymers break down. Plastic polymers with a higher proportion of side chains tend to have more branching and are consequently less susceptible to microbial breakdown. An inverse relationship exists between crystallinity and degradation rate, as non-crystalline segments of polymers are more vulnerable to enzymatic degradation due to their looser packing and accessibility (Devi et al. 2016). Additionally, the biodegradability of MPs can be significantly altered by the presence of plasticizers or additives. Volke-Seplveda et al. (2002) demonstrated how the presence of ethanol influenced the ability of fungi to break down low-density polyethylene (LDPE), with a significant increase in mineralization rates in the presence of ethanol. Similarly, a study by Satlewal et al. (2008) found that microbial consortia were more effective in degrading high-density polyethylene (HDPE) compared to LDPE, indicating that MP biodegradation is influenced by density as well. Jeon and Kim (2013) explored the degradation of polylactic acid (PLA) mediated by *S. maltophilia* LB 2-3 in the presence of UV radiation, observing a rapid decrease in molecular weight and tensile properties of PLA upon UV exposure. Hydrophobicity and hydrophilicity of the polymer surface also play a role in microbe-mediated degradation. Increased hydrophilicity enhances initial microbial colonization of plastic polymers, while high hydrophobicity is thought to impede extracellular enzyme activity. Greater wettability of hydrophilic surfaces, characterized by higher surface energies and lower contact angles with water promotes microbial adhesion to the polymer surface, expediting their breakdown (Chamas et al. 2020).

5.5 MECHANISMS OF MICROBE-MEDIATED MICROPLASTIC DEGRADATION

The microbial degradation of microplastics (MPs) involves a complex interplay of biochemical events, and the breakdown of different types of MPs occurs through various mechanisms. Several factors, including the chemical composition and molecular weight of MPs, the types of microbes present, and environmental conditions, influence the microbe-mediated degradation of MPs. These degradation processes can be categorized as physical, chemical, or biological, with a multitude of enzymes playing crucial roles in the biological degradation process (Padervand et al. 2020; Bacha et al. 2021). The fundamental procedure begins with the fragmentation of larger polymer structures into smaller particles, followed by the breakdown of these smaller polymers into oligomers, dimers, and eventually monomers. Microorganisms facilitate these processes through mineralization. During mineralization, carbon dioxide is released, along with the production of intermediary molecules that serve as an energy source for microbial growth. Several extracellular enzymes are essential for the breakdown of microplastics, including esterases, lipases, lignin peroxidases, laccases, and manganese peroxidases. These enzymes play a critical role in rendering microplastics more hydrophilic and transforming them into carbonyl or alcohol residues. Microbes utilize the tricarboxylic acid cycle and the β-oxidation pathway to break down these carbonyl and hydroxyl groups (Taniguchi et al. 2019). Microorganisms also contribute to intracellular degradation by hydrolyzing endogenous carbon reservoirs. The enzymes produced by microbes are responsible for extracellular degradation at the same time, which results in smaller molecules, such

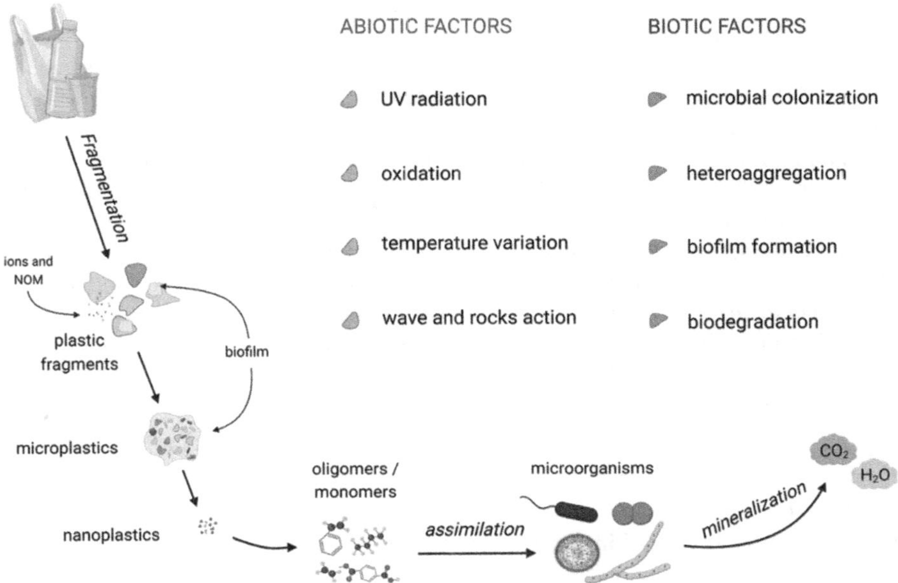

FIGURE 5.2 Graphical representation of microplastic degradation through combined biotic and abiotic factors. (Adapted from Anand et al., 2023, with permission from Springer Nature.)

as oligomers, dimers, and monomers. This procedure is well known as depolymerization. Microorganisms absorb these smaller molecules into their cytoplasm and use them in a variety of metabolic pathways; after that, they are further broken down into stable end products such as carbon dioxide, water, or methane (Figure 5.2).

5.6 ROLE OF FUNGI IN MICROPLASTIC DEGRADATION

Fungi and bacteria play pivotal roles in maintaining biogeochemical cycles and essential nutrients on Earth. Many fungal species have been identified as proficient degraders of various plastic polymers due to their ability to utilize these synthetic plastics as their primary or sole sources of carbon and energy. Fungi, as heterotrophic organisms, acquire nutrients by absorbing them from their surroundings. Pathak (2017) noted that fungi excrete extracellular enzymes from their hyphae, which break down large organic molecules into smaller organic compounds that can be reabsorbed. This enzymatic process results in the release of carbon dioxide (CO_2) and water (H_2O) under aerobic conditions and methane (CH_4) under anaerobic conditions. Fungi employ a range of unique strategies to tackle complex chemicals, including harmful pollutants. These strategies encompass a potent enzymatic system, adsorption capacity, and the production of natural biosurfactants like hydrophobins. These mechanisms enable fungi to utilize polymers as sources of carbon and electrons in their metabolic processes. Among fungal groups, the *Aspergillus* spp.

are particularly noteworthy for their involvement in the biodegradation of synthetic polymers. Various *Aspergillus* species isolated from terrestrial environments, such as *A. fumigatus* (Osman et al. 2018) and *A. clavatus* (Gajendiran et al. 2016), have demonstrated the capability to degrade polyethylene (PE), polyester polyurethane (PU), and polypropylene (PP), respectively. Furthermore, under submerged and solid-state fermentation conditions, endophytic fungi isolated from several plant species have demonstrated variable degrees of PU degradation (Russell et al. 2011). The degradation of several polymers, such as polyethylene, polypropylene, and polyethylene terephthalate, has also been linked to other genera, including *Penicillium simplicissimum*, *Aspergillus niger*, *Zalerion maritimum*, and *Cladosporium* (Paço et al. 2019; Oliveira et al. 2020). The main carbon source for these fungi is microplastics, which are broken down by extracellular enzymes. Few more examples of plastic-degrading fungi are listed in Table 5.1.

TABLE 5.1
Fungal Genera Involved in Plastic Degradation (Published in and after 2015)

Fungi	Plastic Type	Degradation Efficiency	Incubation Period	References
Aspergillus niger, *A. flavus*	Films and granules of LDPE	26% and 16% with *A. niger* and *A. flavus*, respectively	180 days	Deepika and Jaya (2015)
Fusarium oxysporum, *A. fumigatus*, *A. niger*, *Penicillium* spp.	LDPE and PU	*A. niger and F. oxysporum had the highest capacity for degradation*	90 days	Raghavendra et al. (2016)
Aspergillus tubingensis	PU film	100%	60 days	Khan et al. (2017)
P. chrysogenum, *Penecillium oxalicum*	Sheets of HDPE	Morphological changes were observed on subjected sheets	60 days	Ojha et al. (2017)
Monascus ruber, *Monascus* sp.	PU	Between the used fungi, *Monascus* sp. was observed to be much better for the degradation	14 days	El-Morsy et al. (2017)
Aspergillus oryzae	Sheets of LDPE	36.4%–41.93%	4 months	Muhonja et al. (2018)
Trichoderma viride, *A. nominus*	LDPE films	Degradation of 5.13% with *T. viride* and 6.63% with *A. nominus*	45 days	Munir et al. (2018)
Chaetomium globosum	Polyvinyl chloride (PVC)	75%	28 days	Vivi et al. (2019)
Aspergillus terreus, *Chaetomium globosum*	Polyethylene (PE)	58.51% reduction in weight and 94.44% reduction in tensile strength	60 days	Sangale et al. (2019)
Aspergillus flavus	HDPE microplastic particles (MPP)	HDPE MPP was degraded into MPP	28 days	Zhang et al. (2021)

5.7 ROLE OF BACTERIA IN MICROPLASTIC DEGRADATION

Bacteria are known as the engine of the Earth's nutrients. Their function in breakdown, like that of other microbes, is to make sure that nutrients and carbons are released from various complex polymers of both manufactured and natural origins. Various bacterial species have been reported for their critical functions in the bioremediation of microplastic, covering a wide range of materials. It has already been reported that bacterial species belonging to the genera *Pseudomonas*, *Escherichia*, and *Bacillus* are potentially capable of degrading such plastic via applying a range of techniques including metagenomics, cloning, pure culture, and even computational methods. It's interesting to note that these bacteria have also been identified from a variety of biological environments, including dumpsites, recycling facilities, and landfills (Gaytán et al. 2020), cold marine environments, and gut of insects (Ren et al. 2019). It is reported that among various species, *Pseudomonas* is the well-known and thoroughly studied bacterial genus. Such capabilities of breaking down plastic are directly linked with their natural capabilities to break down long-chain fatty acids (Wilkes and Aristilde 2017). Auta et al. (2018) showed in their study that after 40 days of incubation, polypropylene was efficiently broken down by *Bacillus* and *Rhodococcus* bacterial strains with an efficiency of 4.0% and 6.4%, respectively. Further investigations in this direction revealed that microplastics might be broken down by the microorganisms *Bacillus cereus* and *Bacillus gottheilii*. The mechanistic investigation also suggested that biofilm growth favors the bacterial degradation of plastics because it permits colony addition to the surface and prolongs the colony's survival. Although the majority of research is focused on the biodegradability of bacteria as a pure culture, in nature, bacteria typically act in consortia, which have also been shown in various reports (Lwanga et al. 2018; Park and Kim 2019). Few more examples of plastic-degrading bacteria are listed in Table 5.2.

TABLE 5.2
Bacterial Genera Involved in Plastic Degradation (Published in and after 2015)

Bacteria	Plastic Type	Degradation Efficiency	Incubation Period	References
Exiguobacterium sp. strain YT2	Polystyrene	Degradation efficiency reported as 7.4%	60 days	Yang et al. (2015)
Nitrosomonas sp., *Nitrobacter* sp., *Pseudomomas* sp.	Polypropylene, HDPE, and LDPE	15%–20% reduction in HDPE and 5%–9% reduction in polypropylene	90 days	Muenmee et al. (2016)
Pseudomonas aeruginosa and *Escherichia coli*	Polyurethanes	2.5% for *Pseudomonas aeruginosa* and 2.4% for *Escherichia coli*	NA	Uscátegui et al. (2016)
Agios onoufrios	Polystyrene films	0.19%	180 days	Syranidou et al. (2017)
Bacillus licheniformis	Polypropylene and poly-L-lactide	10%	15 days	Jain et al. (2018)

(Continued)

TABLE 5.2 (*Continued*)
Bacterial Genera Involved in Plastic Degradation (Published in and after 2015)

Bacteria	Plastic Type	Degradation Efficiency	Incubation Period	References
Bacillus sp. strain 27	Polypropylene	4%	40 days	Auta et al. (2018)
Bacillus sp. and *Paenobacillus* sp.	Polyethylene	14.7%	60 days	Park and Kim (2019)
Enterobacter sp. nov. bt DSCE02, *Enterobacter cloacae* nov. bt DSCE02 and *Pseudomonas aeruginosa* nov. bt DSCE-CD03	LDPE and polypropylene	64.25% and 63%	160 days	Skariyachan et al. (2021)
Lysinibacillus sp.	Polypropylene and polyethylene	4% and 9%	26 days	Jeon et al. (2021)

5.8 ROLE OF ALGAE IN MICROPLASTIC DEGRADATION

Numerous types of algae, encompassing both heterotrophic and photosynthetic types, have been subject to extensive investigation for their significant contributions to bioremediation. These algae possess the ability to mitigate environmental pollution by acquiring, adsorbing, or metabolizing contaminants, reducing them to safer levels, both for inorganic and organic pollutants (Hwang et al. 2020). Nonetheless, it is worth noting that only a limited number of studies have explored the potential of algae in breaking down plastic polymers. Most research efforts have been directed toward their capacity to generate eco-friendly or "green" polymers. In contrast to bacterial systems, algae do not rely on a rich carbon supply for their growth and can adapt to a wide array of environments, including those where microplastics are predominantly found. Within wastewater streams, microalgae are known to colonize plastic surfaces, leading to the initiation of plastic breakdown through the action of exopolysaccharides and ligninolytic enzymes. These polymers primarily serve as a source of carbon, proteins, and carbohydrates while also promoting the growth of algae. Recent scanning electron microscopy studies have revealed instances of algal colonization leading to the degradation of low-density polyethylene sheets (Sanniyasi et al. 2021). *Anabaena spiroides*, *Navicula pupula*, and *Scenedesmus dimorphus* have all been reported by Kumar et al. (2017) for their capabilities of decomposing low as well as high-density polyethylene. After a 30-day inoculation period, the blue-green algae *A. spiroides* showed a high potential for the degradation of low-density polyethylene with an efficiency of 8.18%. The breakdown of microplastics is mainly associated with the growth of biofilms on polymer surfaces. Potential biofilm producers on microplastic polymers have also been reported as cyanobacterial strains, such as *Microcystis*, *Rivularia*, *Pleurocapsa*, *Synechococcus*, *Prochlorothrix*, *Leptolyngbya*, *Calothrix*, and *Scytonema* (Dussud et al. 2018; Muthukrishnan et al. 2019).

5.9 ROLE OF ACTINOMYCETES IN MICROPLASTIC DEGRADATION

Actinomycetes are a diverse genus of filamentous bacteria that are found in soil, plant tissues, and marine environments. They have a wide variety of biotechnological applications in the domains of medicine, food production, and bioremediation. *Actinomadura* spp., *Rhodococcus ruber*, *Streptomyces* groups, and thermophilic *Thermoactinomyces* species are among the actinomycetes with considerable plastic biodegradative potentials that have been identified from diverse ecological zones (Auta et al. 2018; Jabloune et al. 2020). They are reported to be producing N-acetylglucosamine-rich extracellular polymers such as dextran, glycogen, levan, and slime polysaccharides, which may facilitate their ability to cling to plastic surfaces for microbial activity (Pujic et al. 2015). Additionally, it has been shown that the production of biofilms is crucial for the colonization of polymers by actinomycetes. *Streptomyces scabies*, a bacterium isolated from potatoes, was discovered to degrade polyethylene terephthalate (PET) as well as other polymers, including p-nitrophenyl esters, cutin, and suberin using an esterase enzyme with a wide range of substrate specificity (Jabloune et al. 2020). The effectiveness of plastic degradation has also been shown by a microbial consortium, with a large proportion of actinomycetes species degrading polyurethane and other chemical additives (Gaytán et al. 2020).

5.10 ENGINEERING ENZYMES IN MICROBES FOR MICROPLASTIC DEGRADATION

The advancement of genetic engineering techniques has empowered us to manipulate the genetic composition of microorganisms, thereby enhancing their efficiency in biodegradation processes. To bolster microorganisms' remediation potential in the presence of diverse hydrocarbons and heavy metals, several methods, including recombinant DNA (rDNA) technology, gene cloning, and genetic modifications, have been employed. These techniques are instrumental in forging new metabolic pathways and can alter the selectivity and affinity of enzymes toward various microplastics. The essential prerequisites for effective gene editing entail identifying suitable host species, such as *E. coli*, for gene expression and pointing out the pertinent genes responsible for metabolizing and degrading microplastics. Three primary approaches, namely site-directed mutagenesis, antisense RNA technology, and polymerase chain reaction (PCR), are pivotal in this endeavor. For instance, to enhance the microplastic degradation capability of *Archaeoglobus fulgidus*, Lameh et al. (2022) employed in silico site-directed mutagenesis to modify carboxylesterase, resulting in the creation of butylyne terephthalate-co-adipate (BTA) hydrolase. The critical enzymes responsible for microplastic breakdown, such as manganese-dependent peroxidase, were produced using genetically modified strains of *E. coli* and *S. cerevisiae* BY 4741. Similarly, strains of *E. coli* BL21 and *P. chrysosporium* were genetically modified to produce laccase enzymes through a similar approach (Sharma et al. 2018; Paço et al. 2019). These genetically altered enzymes can degrade polyethylene terephthalate more effectively and retain more than 50% activity after 10 hours at 70°C while showing broad pH stability. Ma et al. (2018) developed novel high-efficiency poly(ethylene terephthalate) hydrolase (PETase) mutants through protein engineering. The mutants exhibited 1.4-fold, 2.1-fold, and 2.5-fold increased PET film degrading ability than the wild-type PETase. Overall, engineering enzymes for microbe-mediated microplastic breakdown have demonstrated significant success.

5.11 CONCLUSION AND FUTURE PERSPECTIVE

Concerns regarding the contamination of microplastics (MPs) have long been prevalent in the realms of environmental engineering and ecological research. Exploring the behavior of MP biodegradation plays a pivotal role in our quest to comprehend how to mitigate the persistence of MPs in the environment and sets the stage for the advancement of cutting-edge technologies aimed at addressing pollution-related challenges. This chapter serves as a summary of recent literature pertaining to the microbial degradation of MPs, shedding light on the mechanisms underpinning this degradation process. Over time, numerous functional microbes capable of degrading MPs have been identified, and various techniques for characterizing them have emerged. However, it's worth noting that many aspects of this process still require further research and deeper understanding. Recent advancements in metagenome analysis and the engineering of previously uncultivated microbe species derived from contaminated environments hold significant promise for the development of innovative bioremediation procedures. These developments are likely to shape the future of microbial-mediated microplastic degradation.

REFERENCES

Alimi, O. S., Fadare, O. O., & Okoffo, E. D. (2021). Microplastics in African ecosystems: Current knowledge, abundance, associated contaminants, techniques, and research needs. *Science of the Total Environment*, *755*, 142422.

Anand, U., Dey, S., Bontempi, E., Ducoli, S., Vethaak, A. D., Dey, A., & Federici, S. (2023). Biotechnological methods to remove microplastics: A review. *Environmental Chemistry Letters*, *21*, 1–24.

Auta, H. S., Emenike, C. U., Jayanthi, B., & Fauziah, S. H. (2018). Growth kinetics and biodeterioration of polypropylene microplastics by *Bacillus* sp. and *Rhodococcus* sp. isolated from mangrove sediment. *Marine Pollution Bulletin*, *127*, 15–21.

Bacha, A. U. R., Nabi, I., & Zhang, L. (2021). Mechanisms and the engineering approaches for the degradation of microplastics. *ACS ES&T Engineering*, *1*(11), 1481–1501.

Bandmann, V., Müller, J. D., Köhler, T., & Homann, U. (2012). Uptake of fluorescent nano beads into BY2-cells involves clathrin-dependent and clathrin-independent endocytosis. *FEBS Letters*, *586*(20), 3626–3632.

Boots, B., Russell, C. W., & Green, D. S. (2019). Effects of microplastics in soil ecosystems: Above and below ground. *Environmental Science & Technology*, *53*(19), 11496–11506.

Cai, Y., Mitrano, D. M., Heuberger, M., Hufenus, R., & Nowack, B. (2020). The origin of microplastic fiber in polyester textiles: The textile production process matters. *Journal of Cleaner Production*, *267*, 121970.

Cao, D., Wang, X., Luo, X., Liu, G., & Zheng, H. (2017). Effects of polystyrene microplastics on the fitness of earthworms in an agricultural soil. In *IOP Conference Series: Earth and Environmental Science* (Vol. 61, No. 1, p. 012148). IOP Publishing.

Chamas, A., Moon, H., Zheng, J., Qiu, Y., Tabassum, T., & Jang, J. H. & Suh, S. (2020). Degradation rates of plastics in the environment. *ACS Sustainable Chemistry & Engineering*, *8*(9), 3494–3511.

Chia, W. Y., Tang, D. Y. Y., Khoo, K. S., Lup, A. N. K., & Chew, K. W. (2020). Nature's fight against plastic pollution: Algae for plastic biodegradation and bioplastics production. *Environmental Science and Ecotechnology*, *4*, 100065.

Cox, K. D., Covernton, G. A., Davies, H. L., Dower, J. F., Juanes, F., & Dudas, S. E. (2019). Human consumption of microplastics. *Environmental Science & Technology*, *53*(12), 7068–7074.

de Souza Machado, A. A., Kloas, W., Zarfl, C., Hempel, S., & Rillig, M. C. (2018). Microplastics as an emerging threat to terrestrial ecosystems. *Global Change Biology*, *24*(4), 1405–1416.

Deepika, S., & Jaya, M. R. (2015). Biodegradation of low density polyethylene by microorganisms from garbage soil. *Journal of Experimental Biology and Agricultural Sciences*, *3*, 1–5.

Devi, R. S., Kannan, V. R., Natarajan, K., Nivas, D., Kannan, K., Chandru, S., & Antony, A. R. (2016). The role of microbes in plastic degradation. *Environmental Waste Management*, *341*, 341–370.

Dong, H., Chen, Y., Wang, J., Zhang, Y., Zhang, P., Li, X., & Zhou, A. (2021). Interactions of microplastics and antibiotic resistance genes and their effects on the aquaculture environments. *Journal of Hazardous Materials*, *403*, 123961.

Dussud, C., Meistertzheim, A. L., Conan, P., Pujo-Pay, M., George, M., Fabre, P., & Ghiglione, J. F. (2018). Evidence of niche partitioning among bacteria living on plastics, organic particles and surrounding seawaters. *Environmental Pollution*, *236*, 807–816.

El-Morsy, E. M., Hassan, H. M., & Ahmed, E. (2017). Biodegradative activities of fungal isolates from plastic contaminated soils. *Mycosphere*, *8*(8), 1071–1087.

Gajendiran, A., Krishnamoorthy, S., & Abraham, J. (2016). Microbial degradation of low-density polyethylene (LDPE) by *Aspergillus clavatus* strain JASK1 isolated from landfill soil. *3 Biotech*, *6*, 1–6.

Gaytán, I., Sánchez-Reyes, A., Burelo, M., Vargas-Suárez, M., Liachko, I., Press, M., & Loza-Tavera, H. (2020). Degradation of recalcitrant polyurethane and xenobiotic additives by a selected landfill microbial community and its biodegradative potential revealed by proximity ligation-based metagenomic analysis. *Frontiers in Microbiology*, *10*, 2986.

Geyer, R., Jambeck, J. R., & Law, K. L. (2017). Production, use, and fate of all plastics ever made. *Science Advances*, *3*(7), e1700782.

Gruber, E. S., Stadlbauer, V., Pichler, V., Resch-Fauster, K., Todorovic, A., Meisel, T. C., & Kenner, L. (2023). To waste or not to waste: Questioning potential health risks of micro-and nanoplastics with a focus on their ingestion and potential carcinogenicity. *Exposure and Health*, *15*(1), 33–51.

He, P., Chen, L., Shao, L., Zhang, H., & Lü, F. (2019). Municipal solid waste (MSW) landfill: A source of microplastics? Evidence of microplastics in landfill leachate. *Water Research*, *159*, 38–45.

Heuchan, S. M., Fan, B., Kowalski, J. J., Gillies, E. R., & Henry, H. A. (2019). Development of fertilizer coatings from polyglyoxylate–polyester blends responsive to root-driven pH change. *Journal of Agricultural and Food Chemistry*, *67*(46), 12720–12729.

Hwang, J. H., Sadmani, A., Lee, S. J., Kim, K. T., & Lee, W. H. (2020). Microalgae: An eco-friendly tool for the treatment of wastewaters for environmental safety. *Bioremediation of Industrial Waste for Environmental Safety: Volume II: Biological Agents and Methods for Industrial Waste Management*, 283–304. https://doi.org/10.1007/978-981-13-3426-9_12.

Jabloune, R., Khalil, M., Moussa, I. E. B., Simao-Beaunoir, A. M., Lerat, S., Brzezinski, R., & Beaulieu, C. (2020). Enzymatic degradation of p-nitrophenyl esters, polyethylene terephthalate, cutin, and suberin by Sub1, a suberinase encoded by the plant pathogen *Streptomyces scabies*. *Microbes and Environments*, *35*(1), ME19086.

Jain, K., Bhunia, H., & Sudhakara Reddy, M. (2018). Degradation of polypropylene–poly-L-lactide blend by bacteria isolated from compost. *Bioremediation Journal*, *22*(3–4), 73–90.

Jambeck, J. R., Geyer, R., Wilcox, C., Siegler, T. R., Perryman, M., Andrady, A., & Law, K. L. (2015). Plastic waste inputs from land into the ocean. *Science, 347*(6223), 768–771.

Jeon, H. J., & Kim, M. N. (2013). Biodegradation of poly (L-lactide) (PLA) exposed to UV irradiation by a mesophilic bacterium. *International Biodeterioration & Biodegradation, 85*, 289–293.

Jeon, J. M., Park, S. J., Choi, T. R., Park, J. H., Yang, Y. H., & Yoon, J. J. (2021). Biodegradation of polyethylene and polypropylene by *Lysinibacillus* species JJY0216 isolated from soil grove. *Polymer Degradation and Stability, 191*, 109662.

Jeong, C. B., Won, E. J., Kang, H. M., Lee, M. C., Hwang, D. S., Hwang, U. K., & Lee, J. S. (2016). Microplastic size-dependent toxicity, oxidative stress induction, and p-JNK and p-p38 activation in the monogonont rotifer (Brachionus koreanus). *Environmental Science & Technology, 50*(16), 8849–8857.

Jiang, X., Chen, H., Liao, Y., Ye, Z., Li, M., & Klobučar, G. (2019). Ecotoxicity and genotoxicity of polystyrene microplastics on higher plant Vicia faba. *Environmental Pollution, 250*, 831–838.

Khalid, N., Aqeel, M., Noman, A., Hashem, M., Mostafa, Y. S., Alhaithloul, H. A. S., & Alghanem, S. M. (2021). Linking effects of microplastics to ecological impacts in marine environments. *Chemosphere, 264*, 128541.

Khan, S., Nadir, S., Shah, Z. U., Shah, A. A., Karunarathna, S. C., Xu, J. & Hasan, F. (2017). Biodegradation of polyester polyurethane by *Aspergillus tubingensis. Environmental Pollution, 225*, 469–480.

Kumar, M., Xiong, X., He, M., Tsang, D. C., Gupta, J., Khan, E., & Bolan, N. S. (2020). Microplastics as pollutants in agricultural soils. *Environmental Pollution, 265*, 114980.

Kumar, R. V., Kanna, G. R., & Elumalai, S. (2017). Biodegradation of polyethylene by green photosynthetic microalgae. *Journal of Bioremediation & Biodegradation, 8*(381), 2.

Lameh, F., Baseer, A. Q., & Ashiru, A. G. (2022). Retracted: Comparative molecular docking and molecular-dynamic simulation of wild-type-and mutant carboxylesterase with BTA-hydrolase for enhanced binding to plastic. *Engineering in Life Sciences, 22*(1), 13–29.

Lei, L., Liu, M., Song, Y., Lu, S., Hu, J., Cao, C., & He, D. (2018). Polystyrene (nano) microplastics cause size-dependent neurotoxicity, oxidative damage and other adverse effects in Caenorhabditis elegans. *Environmental Science: Nano, 5*(8), 2009–2020.

Leslie, H. A., Van Velzen, M. J., Brandsma, S. H., Vethaak, A. D., Garcia-Vallejo, J. J., & Lamoree, M. H. (2022). Discovery and quantification of plastic particle pollution in human blood. *Environment International, 163*, 107199.

Li, L., Luo, Y., Li, R., Zhou, Q., Peijnenburg, W. J., Yin, N., ... & Zhang, Y. (2020). Effective uptake of submicrometre plastics by crop plants via a crack-entry mode. *Nature Sustainability, 3*(11), 929–937.

Lwanga, E. H., Thapa, B., Yang, X., Gertsen, H., Salánki, T., Geissen, V., & Garbeva, P. (2018). Decay of low-density polyethylene by bacteria extracted from earthworm's guts: A potential for soil restoration. *Science of the Total Environment, 624*, 753–757.

Ma, Y., Yao, M., Li, B., Ding, M., He, B., Chen, S., & Yuan, Y. (2018). Enhanced poly (ethylene terephthalate) hydrolase activity by protein engineering. *Engineering, 4*(6), 888–893.

Muenmee, S., Chiemchaisri, W., & Chiemchaisri, C. (2016). Enhancement of biodegradation of plastic wastes via methane oxidation in semi-aerobic landfill. *International Biodeterioration & Biodegradation, 113*, 244–255.

Muhonja, C. N., Makonde, H., Magoma, G., & Imbuga, M. (2018). Biodegradability of polyethylene by bacteria and fungi from Dandora dumpsite Nairobi-Kenya. *PLoS One, 13*(7), e0198446.

Munir, E., Harefa, R. S. M., Priyani, N., & Suryanto, D. (2018). Plastic degrading fungi *Trichoderma viride* and *Aspergillus nomius* isolated from local landfill soil in Medan. In *IOP Conference Series: Earth and Environmental Science* (Vol. 126, No. 1, p. 012145). IOP Publishing.

Muthukrishnan, T., Al Khaburi, M., & Abed, R. M. (2019). Fouling microbial communities on plastics compared with wood and steel: Are they substrate-or location-specific? *Microbial Ecology*, *78*, 361–374.

Nava, V., & Leoni, B. (2021). A critical review of interactions between microplastics, microalgae and aquatic ecosystem function. *Water Research*, *188*, 116476.

Ojha, N., Pradhan, N., Singh, S., Barla, A., Shrivastava, A., Khatua, P. & Bose, S. (2017). Evaluation of HDPE and LDPE degradation by fungus, implemented by statistical optimization. *Scientific Reports*, *7*(1), 39515.

Oliveira, J., Belchior, A., da Silva, V. D., Rotter, A., Petrovski, Ž., Almeida, P. L., & Gaudêncio, S. P. (2020). Marine environmental plastic pollution: Mitigation by microorganism degradation and recycling valorization. *Frontiers in Marine Science*, *7*, 567126.

Osman, M., Satti, S. M., Luqman, A., Hasan, F., Shah, Z., & Shah, A. A. (2018). Degradation of polyester polyurethane by *Aspergillus* sp. strain S45 isolated from soil. *Journal of Polymers and the Environment*, *26*, 301–310.

Paço, A., Jacinto, J., da Costa, J. P., Santos, P. S., Vitorino, R., Duarte, A. C., & Rocha-Santos, T. (2019a). Biotechnological tools for the effective management of plastics in the environment. *Critical Reviews in Environmental Science and Technology*, *49*(5), 410–441.

Padervand, M., Lichtfouse, E., Robert, D., & Wang, C. (2020). Removal of microplastics from the environment. A review. *Environmental Chemistry Letters*, *18*, 807–828.

Park, S. Y., & Kim, C. G. (2019). Biodegradation of micro-polyethylene particles by bacterial colonization of a mixed microbial consortium isolated from a landfill site. *Chemosphere*, *222*, 527–533.

Pathak, V. M. (2017). Review on the current status of polymer degradation: A microbial approach. *Bioresources and Bioprocessing*, *4*, 1–31.

Plastics Europe. (2019). Plastics—the facts. February 4. Retrieved from https://www.plasticseurope.org/application/files/5515/1689/9220/2014plastics_ the_facts_PubFeb2015.pdf.

Pujic, P., Beaman, B. L., Ravalison, M., Boiron, P., & Rodríguez-Nava, V. (2015). Nocardia and Actinomyces. In *Molecular Medical Microbiology* (pp. 731–752). Academic Press. https://doi.org/10.1016/B978-0-12-397169-2.00040-8.

Qi, Y., Yang, X., Pelaez, A. M., Lwanga, E. H., Beriot, N., Gertsen, H. & Geissen, V. (2018). Macro-and micro-plastics in soil-plant system: Effects of plastic mulch film residues on wheat (Triticum aestivum) growth. *Science of the Total Environment*, *645*, 1048–1056.

Raghavendra, V. B., Uzma, M., Govindappa, M., Vasantha, R. A., & Lokesh, S. (2016). Screening and identification of polyurethane (PU) and low density polyethylene (LDPE) degrading soil fungi isolated from municipal solid waste. *International Journal of Current Research*, *8*(7), 34753–34761.

Ren, L., Men, L., Zhang, Z., Guan, F., Tian, J., Wang, B. & Zhang, W. (2019). Biodegradation of polyethylene by *Enterobacter* sp. D1 from the guts of wax moth *Galleria mellonella*. *International Journal of Environmental Research and Public Health*, *16*(11), 1941.

Russell, J. R., Huang, J., Anand, P., Kucera, K., Sandoval, A. G., Dantzler, K. W. & Strobel, S. A. (2011). Biodegradation of polyester polyurethane by endophytic fungi. *Applied and Environmental Microbiology*, *77*(17), 6076–6084.

Sander, M. (2019). Biodegradation of polymeric mulch films in agricultural soils: Concepts, knowledge gaps, and future research directions. *Environmental Science & Technology*, *53*(5), 2304–2315.

Sangale, M. K., Shahnawaz, M., & Ade, A. B. (2019). Potential of fungi isolated from the dumping sites mangrove rhizosphere soil to degrade polythene. *Scientific Reports*, *9*(1), 1–11.

Sanniyasi, E., Gopal, R. K., Gunasekar, D. K., & Raj, P. P. (2021). Biodegradation of low-density polyethylene (LDPE) sheet by microalga, *Uronema africanum* Borge. *Scientific Reports*, *11*(1), 17233.

Satlewal, A., Soni, R., Zaidi, M. G. H., Shouche, Y., & Goel, R. (2008). Comparative biodegradation of HDPE and LDPE using an indigenously developed microbial consortium. *Journal of Microbiology and Biotechnology*, *18*(3), 477–482.

Sharma, B., Dangi, A. K., & Shukla, P. (2018). Contemporary enzyme based technologies for bioremediation: A review. *Journal of Environmental Management*, *210*, 10–22.

Shen, M., Zhu, Y., Zhang, Y., Zeng, G., Wen, X., Yi, H. & Song, B. (2019). Micro (nano) plastics: Unignorable vectors for organisms. *Marine Pollution Bulletin*, *139*, 328–331.

Skariyachan, S., Taskeen, N., Kishore, A. P., Krishna, B. V., & Naidu, G. (2021). Novel consortia of *Enterobacter* and *Pseudomonas* formulated from cow dung exhibited enhanced biodegradation of polyethylene and polypropylene. *Journal of Environmental Management*, *284*, 112030.

Sol, D., Laca, A., Laca, A., & Díaz, M. (2020). Approaching the environmental problem of microplastics: Importance of WWTP treatments. *Science of the Total Environment*, *740*, 140016.

Su, Y., Ashworth, V., Kim, C., Adeleye, A. S., Rolshausen, P., Roper, C. & Jassby, D. (2019). Delivery, uptake, fate, and transport of engineered nanoparticles in plants: A critical review and data analysis. *Environmental Science: Nano*, *6*(8), 2311–2331.

Syranidou, E., Karkanorachaki, K., Amorotti, F., Franchini, M., Repouskou, E., Kaliva, M. & Kalogerakis, N. (2017). Biodegradation of weathered polystyrene films in seawater microcosms. *Scientific Reports*, *7*(1), 17991.

Tang, Y., Liu, Y., Chen, Y., Zhang, W., Zhao, J., He, S. & Yang, Z. (2021). A review: Research progress on microplastic pollutants in aquatic environments. *Science of the Total Environment*, *766*, 142572.

Taniguchi, I., Yoshida, S., Hiraga, K., Miyamoto, K., Kimura, Y., & Oda, K. (2019). Biodegradation of PET: Current status and application aspects. *Acs Catalysis*, *9*(5), 4089–4105.

Uscátegui, Y. L., Arévalo, F. R., Díaz, L. E., Cobo, M. I., & Valero, M. F. (2016). Microbial degradation, cytotoxicity and antibacterial activity of polyurethanes based on modified castor oil and polycaprolactone. *Journal of Biomaterials Science, Polymer Edition*, *27*(18), 1860–1879.

Vethaak, A. D., & Legler, J. (2021). Microplastics and human health. *Science*, *371*(6530), 672–674.

Vivi, V. K., Martins-Franchetti, S. M., & Attili-Angelis, D. (2019). Biodegradation of PCL and PVC: *Chaetomium globosum* (ATCC 16021) activity. *Folia Microbiologica*, *64*, 1–7.

Volke-Sepúlveda, T., Saucedo-Castañeda, G., Gutiérrez-Rojas, M., Manzur, A., & Favela-Torres, E. (2002). Thermally treated low density polyethylene biodegradation by Penicillium pinophilum and Aspergillus niger. *Journal of Applied Polymer Science*, *83*(2), 305–314.

Wan, Y., Wu, C., Xue, Q., & Hui, X. (2019). Effects of plastic contamination on water evaporation and desiccation cracking in soil. *Science of the Total Environment*, *654*, 576–582.

Weithmann, N., Möller, J. N., Löder, M. G., Piehl, S., Laforsch, C., & Freitag, R. (2018). Organic fertilizer as a vehicle for the entry of microplastic into the environment. *Science Advances*, *4*(4), eaap8060.

Wilkes, R. A., & Aristilde, L. (2017). Degradation and metabolism of synthetic plastics and associated products by *Pseudomonas* sp.: Capabilities and challenges. *Journal of Applied Microbiology*, *123*(3), 582–593.

Yang, L., Zhang, Y., Kang, S., Wang, Z., & Wu, C. (2021). Microplastics in freshwater sediment: A review on methods, occurrence, and sources. *Science of the Total Environment*, *754*, 141948.

Yang, Y., Yang, J., Wu, W. M., Zhao, J., Song, Y., Gao, L. & Jiang, L. (2015). Biodegradation and mineralization of polystyrene by plastic-eating mealworms: Part 2. Role of gut microorganisms. *Environmental Science & Technology*, *49*(20), 12087–12093.

Zhang, G. S., & Liu, Y. F. (2018). The distribution of microplastics in soil aggregate fractions in southwestern China. *Science of the Total Environment*, *642*, 12–20.

Zhang, J., Gao, D., Li, Q., Zhao, Y., Li, L., Lin, H. & Zhao, Y. (2020). Biodegradation of polyethylene microplastic particles by the fungus *Aspergillus flavus* from the guts of wax moth *Galleria mellonella*. *Science of the Total Environment*, *704*, 135931.

Zhang, S., Wang, J., Yan, P., Hao, X., Xu, B., Wang, W., & Aurangzeib, M. (2021). Non-biodegradable microplastics in soils: A brief review and challenge. *Journal of Hazardous Materials*, *409*, 124525.

Zhou, J., Gui, H., Banfield, C. C., Wen, Y., Zang, H., Dippold, M. A. & Jones, D. L. (2021a). The microplastisphere: Biodegradable microplastics addition alters soil microbial community structure and function. *Soil Biology and Biochemistry*, *156*, 108211.

Zhou, J., Wen, Y., Marshall, M. R., Zhao, J., Gui, H., Yang, Y. & Zang, H. (2021b). Microplastics as an emerging threat to plant and soil health in agroecosystems. *Science of the Total Environment*, *787*, 147444.

Zhu, D., Chen, Q. L., An, X. L., Yang, X. R., Christie, P., Ke, X. & Zhu, Y. G. (2018). Exposure of soil collembolans to microplastics perturbs their gut microbiota and alters their isotopic composition. *Soil Biology and Biochemistry*, *116*, 302–310.

Zurier, H. S., & Goddard, J. M. (2021). Biodegradation of microplastics in food and agriculture. *Current Opinion in Food Science*, *37*, 37–44.

6 Microplastics in Aquaculture
Implication on Fish and Human Health

Amit Ranjan, Kalaiselvan Pandi, and Kavitha Malarvizhi

6.1 INTRODUCTION

Microplastics are small plastic particles less than 5 mm in size that can enter aquatic environments through various sources such as wastewater, agricultural runoff, breakdown of fishing gear and atmospheric deposition (Cole et al., 2011). They can also be introduced through plastic feed bags, plastic-based aquaculture equipment, and microplastic-contaminated water. Fish and other aquatic organisms can ingest these microplastics, causing physical damage and leading to chemical and biological changes in their bodies. This ingestion of microplastics can have adverse effects on their health and growth. Microplastics have detrimental effects on fish, causing physical damage and behavioural changes. They can accumulate in the tissues of farmed fish and shellfish, posing a potential risk to human health if consumed. Studies have found microplastics in the gut contents of various farmed fish species, such as salmon, trout, and tilapia (Smith et al., 2018). Additionally, microplastics in aquaculture systems can accumulate in the surrounding environment, posing risks to other wildlife and ecosystems (Akhbarizadeh et al., 2018).

Microplastics have significant impacts on aquatic species, including fish, crustaceans, mollusks, and oysters. They can affect behaviour, growth, and reproductive success. Aquatic organisms can experience negative effects on growth, development, and overall health, as well as changes in gene expression and immune function. These impacts can ultimately affect the quality and safety of harvested seafood products (Sussarellu et al., 2016). Microplastics also serve as carriers for pollutants, such as persistent organic pollutants like polychlorinated biphenyls (PCBs) and heavy metals such as lead and mercury, which can accumulate in the tissues of aquatic organisms. This accumulation poses a potential risk to human health when consuming seafood (Wang et al., 2021). Studies indicate that the presence of microplastics in fish feed can disturb the normal function of digestive enzymes and antioxidant defences. This disruption can subsequently result in oxidative stress and hindered growth rates among the fish (Li et al., 2021).

DOI: 10.1201/9781003438793-6

In aquaculture systems, microplastic particles are mistakenly perceived as food by fish, leading to their ingestion. Fishmeal, a common ingredient in aquaculture feeds, has been identified as a major pathway for microplastic entry (Ory et al., 2021). When ingested, microplastics can cause false satiation, obstruct the gastrointestinal tract, impair nutrient utilisation, and hinder fish growth (Wang et al., 2021). Concerns have been raised about the presence of microplastics in fish meal, with studies showing varying levels of contamination depending on the source. Fish meal has been found to contain microplastics ranging from 17 to 150 particles per gram, raising concerns about their entry into the food chain through aquaculture (Ma et al., 2021). Studies on various fish species have demonstrated that exposure to microplastics in their feed can result in decreased growth rates, altered gene expression related to the immune system, decreased antioxidant activity, and increased lipid accumulation in their tissues (Niu et al., 2022). These findings suggest that the use of fish meal in aquaculture feeds could contribute significantly to microplastic pollution in the aquatic environment.

Feeding aquatic organisms with fish meal contaminated with microplastics can result in the accumulation of microplastics in their tissues, potentially causing adverse health effects. Shrimp fed with microplastic-contaminated feed exhibit reduced feeding rates, lower weight gain, and decreased survival rates compared to those fed with uncontaminated feed (Pereira et al., 2020). Exposure to microplastics in shrimp also leads to changes in gene expression related to immune response and stress. Microplastics in feed can reduce feeding efficiency and growth rates in shrimp, as well as disrupt their gut microbiome and immune system function. This emphasises the need for careful sourcing and testing of feed ingredients in aquaculture to minimise the risk of microplastic contamination (Wang et al., 2021). Efforts to address the presence of microplastics in fish meal include improving waste management practices, exploring alternative feed ingredients, and developing technologies to filter microplastics from fish meal before its use in aquaculture operations (Lee et al., 2021).

6.2 ENDOGENOUS AND EXOGENOUS ROUTE OF ENTRY OF MICROPLASTICS IN AQUACULTURE SYSTEM

Plastic contamination is a pervasive problem that spans various environments, from the Arctic seafloor to the upper atmosphere (Allen et al., 2021). Microplastics categorised as primary or secondary. Primary microplastics originate from small plastic manufacturing industries, while secondary microplastics are formed when larger plastic fragments degrade. Aquatic ecosystem sediments serve as the primary destination for most microplastics. Major contributors to the presence of microplastics in the ocean include the import of plastic waste from land, the breakdown of marine plastic debris, floating equipment used in aquaculture, and plastic waste from maritime vessels. The origin of freshwater microplastics, in comparison to marine microplastics, is still uncertain. In aquaculture, primary sources of microplastics include industrial sewage runoff, washing wastewater, river inputs, fisheries and aquaculture activities, and tourism development. Additionally, the introduction of microplastics into freshwater ecosystems is attributed to various sources such as industrial raw materials, personal care products, plastic waste in urban areas, synthetic textiles, and plastic waste in tourism and industrial regions (Wright et al., 2020). Microplastics in aquatic environments stem from degraded plastic products like fragments, granules, and films, which originate from fishing gears, plastic bags, containers, bottles, and similar items (Shim et al., 2018).

Aquaculture is impacted by contamination primarily because of the extensive use of plastics such as polypropylene (PP) and polyethylene (PE). Microplastics enter aquatic regions through surface runoff during the rainy season (Ma et al., 2021). Proximity to areas with high human activities leads to elevated levels of microplastics in aquaculture areas. Rivers play a significant role in transferring microplastics from land to freshwater aquaculture ponds and marine environments (Lebreton et al., 2017). The distribution of microplastics in aquaculture environments is influenced by factors such as geography and weather conditions, as reported by Wang et al. (2021). Their study found that typhoons increased the concentration of microplastics in the Gulf region, particularly impacting cultured oysters. Despite efforts to remove microplastics through wastewater treatment plants, the discharged effluent still poses a threat to the aquatic environment. Improper disposal of plastic waste by humans, as well as industrial and agricultural discharges, contributes to soil contamination and water pollution. As a result, microplastics are absorbed by plants, including edible algae and aquatic plants, and enter the food chain, ultimately being present in the bodies of humans and animals.

Improper disposal of plastic products used in fishing and aquaculture, such as bags, bottles, packaging, gear, buoyant materials, and net cages, poses a significant threat to the aquatic environment. Over time, these plastics degrade into microplastics due to prolonged exposure and various forms of wear and tear. Fishing gear made of plastic materials can release microplastics into the aquatic environment through long-term use and inadequate maintenance (Di et al., 2019). Fishing equipment and fish feed are major sources of microplastic release in aquaculture. Microplastics are also present in fish meal, serving as a primary pathway for their entry into the aquaculture environment. Fish offal, which is consumed by fish and other organisms, sustains the presence of microplastics in the food web. Both natural and artificial fish feeds contribute to the release of microplastics during production, storage, and transportation. Fish feed additives, medications, and health-promoting products are additional sources of microplastics (Zhou et al., 2021). Microplastics can be ingested and absorbed by various aquatic organisms, including phytoplankton, zooplankton, small fish, crustaceans, and aquatic plants. This leads to their incorporation into the diet of small fish, which can then be used as feed, live bait, or a natural food source by other organisms in the aquatic environment. Microplastics can accumulate in higher trophic levels, including copepods, mussels, and higher vertebrates, through the food chain. They also provide a suitable surface for the growth of aquatic microorganisms and can be mistaken for natural food, potentially causing adverse effects on human health and aquatic animals (Niu et al., 2022).

In the context of shellfish culture, Bendell (2015) observed that antipredator nets used in the industry actually become a source of microplastics instead of protecting shellfish from predators. A study by Wang et al. (2018) investigated the presence of microplastics in shrimp ponds and found that shrimp were exposed to microplastics through their feed and the surrounding water. The study concluded that microplastics can accumulate in the tissues of farmed shrimp, posing a potential risk to both the environment and human consumers. Su et al. (2021) investigated the presence of microplastics in shrimp farming systems and found significant levels of microplastic particles in both water and shrimp samples. The researchers identified various types of microplastics, including fibres, fragments, and films, indicating multiple potential sources of contamination within the aquaculture environment. Similarly, research by Zhang et al. (2020) examined the transfer of microplastics from feed to shrimp in aquaculture settings. The study revealed that microplastic particles present in the feed were ingested by shrimp, resulting in their accumulation within the shrimp's digestive system.

6.3 THE IMPLICATION OF MICROPLASTICS ON GROWTH, SURVIVAL, AND REPRODUCTIVE EFFICACY OF FISH

The continuous increase in plastic production has caused widespread plastic pollution in the environment and natural water resources. Plastic precursors, such as bisphenol A (BPA), further contribute to this pollution. BPA is commonly used in the production of various plastic materials and has become one of the most abundant chemical compounds in the environment. Studies have shown that continuous exposure to BPA can lead to its bioaccumulation in humans and other organisms through inhalation and ingestion. BPA is released into the aquatic environment through sources such as industrial effluents, landfill leachate, sewage waters, fishing gear, and plastic fishing accessories (Rochester et al., 2013). BPA has been detected not only in natural waters but also in the muscle tissue of various finfish species, including mullet, and seabass from aquaculture pond (Minaz et al., 2022). The presence of these microparticles in both the environment and fish tissues has been correlated with detrimental effects on various factors such as growth, survival, behaviour, fish populations, morphology, and histological structure. Fish can accumulate toxic BPA even at low concentrations, and it plays a significant role in both aquatic and terrestrial food webs. Microplastics have been observed to impede feeding and growth in sea urchins and disrupt their homeostasis. This disruption leads to increased energy expenditure and reduced growth rates in blue mussels. The consumption of microplastics has been shown to obstruct the gastrointestinal tract in fish, leading to lower feeding rates, reduced appetite, and impaired growth (Bhuyan, 2022). Several studies have reported the adverse effects of microplastics, including mortality before reaching maturity, physical impairment in certain fish species, decreased growth and weight, and negative impacts on gastrointestinal physiology, behaviour, growth, and reproductive success (Feng et al., 2021).

According to Naidoo and Glassom (2019), fish treated with plastic showed adverse effects on growth, with significantly lower body depth and length. The accumulation of microplastics has been associated with various negative effects, such as growth reduction, oxidative stress, hormone disruption, and neurotoxicity (Choi et al., 2018). Several studies have examined the growth of teleost fish when exposed to bisphenol A (BPA). Abdel-Tawwab and Hamed (2018) observed reduced growth and feed intake in Nile tilapia (*Oreochromis niloticus*) exposed to BPA, with the fish consuming less feed compared to the control group in a dose-response manner. Pastva et al.

(2001) suggested that the reduced growth and feed intake may be attributed to the disruption of biological processes caused by the detrimental effects of BPA. It was also found that the reduced growth in fish was accompanied by the suppression of genes related to growth hormone and its receptor.

The adverse effects of microplastics on the reproductive systems of aquatic animals are well-documented. BPA, known for its endocrine-disrupting properties, can cause cellular damage in tissues and reproductive abnormalities (Sussarellu et al., 2016). Exposure to polystyrene (PS) microplastics has been shown to reduce reproductive efficacy and growth rates in oysters. BPA exposure in goldfish results in diminished ovarian maturity, but it can be restored upon withdrawal of exposure. Transgenerational effects of BPA have been observed in zebrafish, with impacts on reproductive tissues and fecundity in subsequent generations. BPA exposure in rainbow trout delays embryonic development stages (Akhter et al., 2018). Increased follicles and oocyte atresia have been observed in common carp, zebrafish, and *Catla catla* due to BPA exposure (Migliaccio et al., 2018). BPA exposure reduces semen quality in brown trout, impairs sperm motility and velocity in goldfish, and affects germ cell progression in fathead minnow. BPA negatively impacts sperm function, embryonic development, and fertilisation in various species. The presence of BPA in the environment not only affects the development and reproduction of exposed individuals but also has effects on their offspring, including enhanced oestrogen production and changes in vitellogenin levels in zebrafish. Exposure to bisphenol A (BPA) in the environment has detrimental effects not only on the development and reproduction of exposed individuals but also on their offspring. In zebrafish, it has been found that BPA exposure enhances oestrogen production while reducing the number of spermatogonia and spermatocytes. Additionally, there is a significant change in vitellogenin levels in zebrafish exposed to BPA (Yang et al., 2021).

The presence of microplastics in aquaculture systems has implications for the growth, survival, and reproductive efficacy of cultured fish and shellfish. To address the implications of microplastics on growth, survival, and reproductive efficacy in aquaculture, it is crucial to implement effective waste management practices, develop alternative feed sources free from microplastics, and promote sustainable aquaculture systems. Continued research and monitoring are also necessary to better understand the extent of the impact and develop mitigation strategies to ensure the health, productivity, and sustainability of aquaculture operations.

6.4 PHYSIO-METABOLIC CHANGES ASSOCIATED WITH MICROPLASTICS

Microplastics in aquaculture can induce various physio-metabolic changes in aquatic organisms. These physio-metabolic changes associated with microplastic contamination pose significant challenges to the aquaculture industry. These changes can have significant impacts on the growth and development of cultured organisms. The physiological effects of microplastics on fish can manifest in various ways. These effects can be observed through changes in different physiological parameters. Microplastic ingestion can affect fish feeding behaviour, leading to reduced appetite or changes in feeding patterns (Wright et al., 2020). Microplastic exposure has negative effects on fish, including hindered growth and development, reduced fertility, abnormal gonadal development, changes in reproductive behaviour, oxidative stress, and immune system dysregulation (Rochman et al., 2016). These impacts can result in reduced body size, altered developmental processes, imbalanced reactive oxygen species and antioxidants and impaired defence mechanisms including compromised immune cell activity, antibody production, and response in fish (Lee et al., 2021).

To assess the effects of microplastics on fish, several physiological parameters can be measured. These parameters provide insights into various aspects of fish health and well-being. Here are some examples:

- Biomarkers of Oxidative Stress: Measurements of antioxidant enzyme activities (e.g., superoxide dismutase, catalase) and levels of reactive oxygen species can indicate the level of oxidative stress induced by microplastic exposure (Hossain et al., 2020).
- Immunological Responses: Evaluation of immune parameters such as immune cell counts, antibody production, and cytokine levels can help determine the impact of microplastics on fish immune function (Rist et al., 2017).
- Haematological Parameters: Analysis of blood parameters including haematocrit, haemoglobin concentration, red blood cell count, and white blood cell count can provide insights into the physiological response of fish to microplastic exposure (Hossain et al., 2020).
- Endocrine Disruption: Measurement of hormone levels, such as cortisol, oestradiol, and testosterone, can help assess the potential endocrine-disrupting effects of microplastics on fish reproductive and physiological processes (Karami et al., 2018).
- Histopathological Examination: Microscopic analysis of fish tissues can reveal any abnormal cellular structures or tissue damage caused by microplastic exposure (Rochman et al., 2016).

Microplastics cause physical damage to the inner linings of the gut, leading to ulceration and damage to the villi morphology and functionality. They affect the epithelial cells, such as cryptic glandular and goblet cells, and result in clogging of blood vessels and infiltration of white blood cells. Microplastics also have adverse effects on the liver, where they cause single cell necrosis, glycogen deficiency, lipid deprivation, and hepatotoxicity (Figure 6.1). The liver serves as a permanent residence for microplastics after absorption, unlike in the gastrointestinal tract.

Bioaccumulation and bioconcentration of microplastics in aquaculture can have adverse effects on the health and well-being of the cultured organisms. The accumulated microplastics can cause physical damage, oxidative stress, immune system disruptions, and other physiological and metabolic changes. Furthermore, if these organisms are consumed by humans or other animals, there is a potential for the transfer of microplastics up the food chain, leading to wider ecological and health implications.

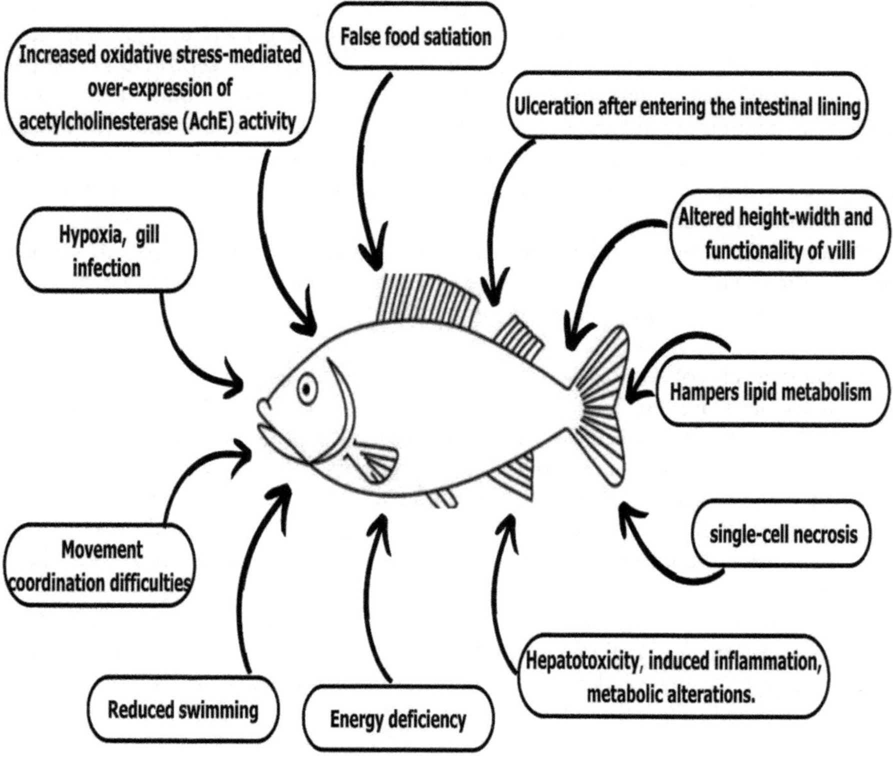

FIGURE 6.1 Potential impacts of microplastics on fish physiology.

6.5 HUMAN HEALTH IMPLICATIONS OF MICROPLASTICS ASSOCIATED WITH AQUACULTURE PRODUCE

The potential effects of microplastics associated with aquaculture produce on human health are of concern. The presence of microplastics in aquaculture produce raises questions about the potential ingestion and exposure of these particles to consumers. Understanding the human health implications involves investigating the possible transfer of microplastics and associated chemicals from aquaculture organisms to humans, as well as the potential health risks associated with their ingestion. Assessing the safety and long-term impacts of microplastics in aquaculture produce is essential for protecting human health and ensuring the sustainability of the aquaculture industry. It is crucial to monitor and regulate the use of plastic materials in aquaculture practices to minimise the presence of microplastics in aquaculture produce and protect human health. Aquaculture products, such as fish and shellfish, have been discovered to harbour microplastics, and consuming them may entail health risks like inflammation, oxidative stress, and potential carcinogenic effects (Rochman et al., 2016). A study conducted by Su et al. (2021) found that microplastics from aquaculture farms can accumulate in the gut of marine animals, which can

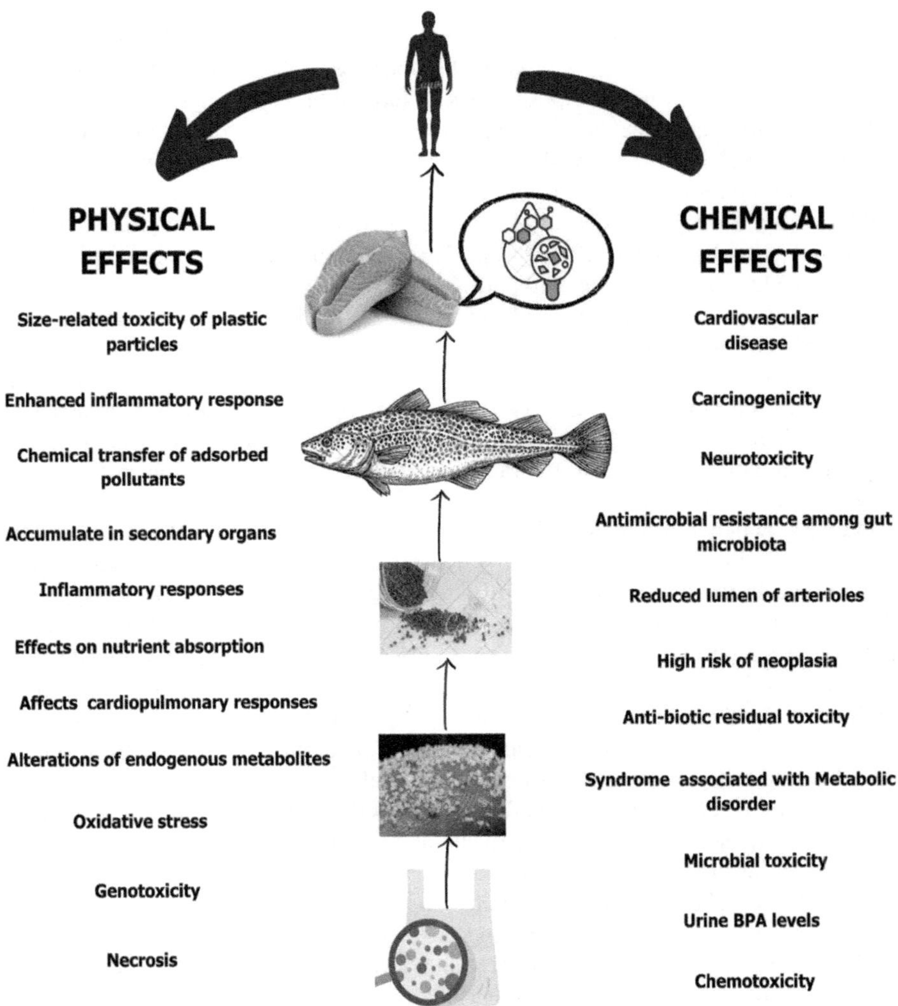

FIGURE 6.2 Human health implications of microplastics associated with aquaculture produce.

then be transferred to humans who consume these animals. The study reported that microplastics in fish and shellfish can cause adverse health effects, including endocrine disruption and immune dysfunction (Figure 6.2). Microplastics can also act as a carrier for other harmful contaminants such as heavy metals and persistent organic pollutants, which can accumulate in the bodies of fish and shellfish and subsequently for humans upon consumption (Wu et al., 2021). Microplastics can absorb and transport toxins and pollutants, which can be transferred to humans upon consumption of contaminated seafood (Li et al., 2018).

Microplastics in seafood may cause damage to human intestinal cells, which could lead to inflammation and other health issues (Yang et al., 2021). Microplastics found in seafood have the potential to harbour toxic chemicals, which can accumulate in seafood tissues and pose a risk to human health if consumed (Cai et al., 2020). Research indicates that microplastics can induce harm to human organs, such as the liver and intestines, while also disrupting the gut microbiome, potentially resulting in inflammation and various health complications (Boucher & Friot, 2017). Additionally, microplastics have been found to contain harmful chemicals, such as bisphenol A and phthalates, which can leach into the food and potentially cause adverse health effects (Huang et al., 2018).

A recent study conducted by Ho et al. (2021) analysed microplastics in three commonly consumed aquaculture products: fish, shrimp, and bivalves. The research revealed that all three products contained microplastics, with bivalves showing the highest concentration. The authors suggest that the presence of microplastics in these products is a cause for concern and highlight the need for increased monitoring and regulation of aquaculture practices. Additionally, microplastics may contain or absorb harmful chemicals, such as endocrine disruptors and persistent organic pollutants, which can have long-term health effects on humans (Boucher & Friot, 2017). According to a study by Mattsson et al. (2021), microplastics in seafood can accumulate in the human body, leading to potential health risks. The study found that microplastics could transfer across the gut barrier and reach vital organs, causing inflammation and damage to tissues. This study emphasises the importance of addressing microplastic pollution in the aquatic environment and its implications on human health. A study by Zhu et al. (2020) found that microplastics were present in the gut of farmed fish and shrimp, with concentrations ranging from 0.18 to 8.6 microplastics per gram of tissue. The study also found that microplastics in the digestive system of the seafood could be transferred to human consumers, potentially leading to exposure to harmful chemicals and toxins. Li et al. (2021) conducted a study revealing that the consumption of microplastics by humans can result in the release of toxic chemicals like phthalates and BPA into the body. These chemicals are known as endocrine disruptors and have been associated with various health problems, including cancer, reproductive and developmental issues, and neurological disorders. Another study by Prata et al. (2020) investigated the impact of microplastics on human health and concluded that the immune system is ineffective at removing microplastics, leading to chronic inflammation and an elevated risk of neoplasia.

Apart from direct health effects, microplastics in seafood can indirectly affect human health by serving as carriers for pathogenic bacteria and viruses. In a study by Wu et al. (2021), it was revealed that microplastics have the capacity to serve as hosts for bacteria, potentially amplifying the risk of foodborne illnesses. This finding emphasises the need for greater awareness and management of microplastic contamination in the context of food safety.

6.6 CONCLUSIONS

In conclusion, the implications of microplastics in aquaculture are significant and multifaceted. The presence of microplastics in aquaculture systems poses risks to both aquatic organisms and human consumers. It can lead to physiological and metabolic changes in cultured organisms, affecting growth, development, and reproductive performance. Moreover, the incorporation of microplastic-contaminated fish meal in aquaculture feeds raises concerns about the potential transfer of microplastics up the food chain. The accumulation of microplastics in the digestive tract and liver of aquatic organisms can cause physical damage, inflammation, and hepatotoxicity. Additionally, the non-polar nature of plastics (hydrophobic nature due to hydrocarbon chains) and the presence of endocrine-disrupting compounds in microplastics raise concerns about their potential impacts on the endocrine system and overall health. To mitigate these risks, efforts are needed to reduce microplastic pollution at its source, develop alternative feed sources, and implement sustainable practices in aquaculture. Continued research and monitoring are crucial to further understand the extent and consequences of microplastics in aquaculture and to ensure the safety and sustainability of aquaculture produce. Addressing these implications of microplastics in aquaculture requires a comprehensive approach involving collaboration between industry stakeholders, policymakers, scientists, and consumers to promote sustainable practices and minimise the introduction and spread of microplastics in aquaculture systems.

REFERENCES

Abdel-Tawwab, M., & Hamed, H. S. (2018). Effect of bisphenol A toxicity on growth performance, biochemical variables, and oxidative stress biomarkers of Nile tilapia, *Oreochromis niloticus* (L.). *Journal of Applied Ichthyology*, 34(5), 1117–1125.

Akhbarizadeh, R., Moore, F., Keshavarzi, B., & Amini, N. (2018). Microplastics in different tissues of fish and prawn from the Musa Estuary, Persian Gulf. *Chemosphere*, 205, 80–87. https://doi.org/10.1016/j.chemosphere.2018.04.135.

Akhter, A., Rahaman, M., Suzuki, R. T., Murono, Y., & Tokumoto, T. (2018). Next-generation and further transgenerational effects of bisphenol A on zebrafish reproductive tissues. *Heliyon*, 4(9), e00788.

Allen, S., Allen, D., Baladima, F., Phoenix, V. R., Thomas, J. L., Le Roux, G., & Sonke, J. E. (2021). Evidence of free tropospheric and long-range transport of microplastic at Pic du Midi Observatory. *Nature Communications*, 12(1), 7242.

Bendell, L. I. (2015). Favored use of anti-predator netting (APN) applied for the farming of clams leads to little benefits to industry while increasing nearshore impacts and plastics pollution. *Marine Pollution Bulletin*, 91(1), 22–28.

Bhuyan, M. (2022). Effects of microplastics on fish and in human health. *Frontiers in Environmental Science*, 250. https://doi.org/10.3389/fenvs.2022.827289.

Boucher, J., & Friot, D. (2017). *Primary microplastics in the oceans: A global evaluation of sources*. Gland, Switzerland: International Union for Conservation of Nature.

Cai, H., Xu, M., Li, X., & Shi, H. (2020). Microplastics in commercial fish feeds and their potential effects on aquatic animals. *Journal of Agricultural and Food Chemistry*, 68(43), 12088–12094. https://doi.org/10.1021/acs.jafc.0c05663.

Choi, J. S., Jung, Y. J., Hong, N. H., Hong, S. H., & Park, J. W. (2018). Toxicological effects of irregularly shaped and spherical microplastics in a marine teleost, the sheepshead minnow (*Cyprinodon variegatus*). *Marine Pollution Bulletin*, 129(1), 231–240.

Cole, M., Lindeque, P., Halsband, C., & Galloway, T. S. (2011). Microplastics as contaminants in the marine environment: A review. *Marine Pollution Bulletin*, 62, 2588–2597. https://doi.org/10.1016/j.marpolbul.2011.09.025.

Di, M., Liu, X., Wang, W., & Wang, J. (2019). Manuscript prepared for submission to Environmental Toxicology and Pharmacology pollution in drinking water source areas: Microplastics in the Danjiangkou Reservoir, China. *Environmental Toxicology and Pharmacology*, 65, 82–89.

Feng, S., Zeng, Y., Cai, Z., Wu, J., Chan, L. L., Zhu, J., & Zhou, J. (2021). Polystyrene microplastics alter the intestinal microbiota function and the hepatic metabolism status in marine medaka (*Oryzias melastigma*). *Science of the Total Environment*, 759, 143558.

Ho, S. S., Zou, S., Leung, C. W., Tsui, M. M., Chow, A. T., & Cheung, S. G. (2021). Microplastics in three commonly consumed aquaculture products: Presence and potential risk to human health. *Science of the Total Environment*, 757, 143972. https://doi.org/10.1016/j.scitotenv.2020.143972.

Hossain, M. S., O'Reilly, L. C., Webster, T. F., Wood, R. J., & Rotchell, J. M. (2020). Aquatic toxicity of microplastics and nanoplastics: Lessons learned from pharmaceuticals and their implications for environmental effects. *Environmental Science & Technology*, 54(9), 5375–5390. https://doi.org/10.1021/acs.est.9b06871.

Huang, Y., Li, D., Chen, Q., Yao, Q., Liao, C., Zhu, H., … & Yang, X. (2018). Microplastics in freshwater and food samples collected from the Pearl River in Guangzhou, China. *Environmental Science and Pollution Research*, 25(19), 18781–18792. https://doi.org/10.1007/s11356-018-2142-0.

Karami, A., Golieskardi, A., Choo, C. K., Larat, V., & Galloway, T. S. (2018). The presence of microplastics in commercial salts from different countries. *Scientific Reports*, 8(1), 1–9. https://doi.org/10.1038/s41598-018-19178-6.

Lebreton, L. C., Van Der Zwet, J., Damsteeg, J. W., Slat, B., Andrady, A., & Reisser, J. (2017). River plastic emissions to the world's oceans. *Nature Communications*, 8(1), 15611.

Lee, J. H., Kim, J. H., Kim, K. J., & An, J. (2021). Effects of microplastic contamination on the physiology of farmed aquatic animals and risk reduction strategies. *Journal of Hazardous Materials*, 416, 125905. https://doi.org/10.1016/j.jhazmat.2021.125905.

Li, J., Li, S., Liu, J., & Cai, L. (2021). Health implications of microplastics in the food chain. *Food Control*, 128, 108217. https://doi.org/10.1016/j.foodcont.2021.108217.

Li, J., Yang, D., Li, L., Jabeen, K., & Shi, H. (2018). Microplastics in commercial bivalves from China. *Environmental Pollution*, 237, 448–455. https://doi.org/10.1016/j.envpol.2017.12.008.

Ma, F., Li, D., & Zhang, X. (2021). Microplastics in aquaculture: A review. *Science of the Total Environment*, 787, 147511. https://doi.org/10.1016/j.scitotenv.2021.147511.

Mattsson, K., Johnson, E. V., Malmendal, A., Linse, S., Hansson, L. A., Cedervall, T., & Hansson, L. A. (2021). Brain damage and behavioural disorders in fish induced by plastic nanoparticles delivered through the food chain. *Scientific Reports*, 11(1), 1–14.

Migliaccio, M., Chioccarelli, T., Ambrosino, C., Suglia, A., Manfrevola, F., Carnevali, O., … & Cobellis, G. (2018). Characterization of follicular atresia responsive to BPA in zebrafish by morphometric analysis of follicular stage progression. *International Journal of Endocrinology*, 2018. https://doi.org/10.1155/2018/4298195.eCollection2018.

Minaz, M., Er, A., Ak, K., Nane, İ. D., İpek, Z. Z., Kurtoğlu, İ. Z., & Kayış, Ş. (2022). Short-term exposure to bisphenol A (BPA) as a plastic precursor: Hematological and behavioral effects on Oncorhynchus mykiss and Vimba. *Water, Air, & Soil Pollution*, 233(4), 122.

Naidoo, T., & Glassom, D. (2019). Decreased growth and survival in small juvenile fish, after chronic exposure to environmentally relevant concentrations of microplastic. *Marine Pollution Bulletin*, 145, 254–259.

Niu, Y., Wang, Z., Ma, Y., Luo, W., Gao, L., & Yang, Y. (2022). Microplastics in fish feed impair growth performance, antioxidant activity, and lipid metabolism in Nile tilapia (Oreochromis niloticus). Aquaculture, 546, 737310. https://doi.org/10.1016/j.aquaculture.2021.737310.

Ory, N., Mazurais, D., Severe, A., Frederich, B., Santoul, F., & Viricel, A. (2021). Microplastic contamination in European sea bass and pangasius catfish feeds and its effect on growth, diet efficiency and health status. Aquaculture, 534, 736438. https://doi.org/10.1016/j.aquaculture.2021.736438.

Pastva, S. D., Villalobos, S. A., Kannan, K., & Giesy, J. P. (2001). Morphological effects of Bisphenol-A on the early life stages of medaka (*Oryzias latipes*). *Chemosphere*, 45(4–5), 535–541.

Pereira, C. D. S., Gonçalves, F. D., Neves, R. F., & Cavalli, R. O. (2020). Exposure to microplastics induce changes on gene expression in Pacific white shrimp Litopenaeus vannamei. *Aquatic Toxicology*, 227, 105602. https://doi.org/10.1016/j.aquatox.2020.105602.

Prata, J. C., da Costa, J. P., Lopes, I., Duarte, A. C., & Rocha-Santos, T. (2020). Environmental exposure to microplastics: An overview on possible human health effects. *Science of the Total Environment*, 702, 134455. https://doi.org/10.1016/j.scitotenv.2019.134455.

Rist, S., Baun, A., & Hartmann, N. B. (2017). Ingestion of microplastics by fish and its potential consequences from a physical perspective. *Integrated Environmental Assessment and Management*, 13(3), 510–515. https://doi.org/10.1002/ieam.1829.

Rochester, J. R. (2013). Bisphenol A and human health: A review of the literature. *Reproductive Toxicology*, 42, 132–155.

Rochman, C. M., Tahir, A., Williams, S. L., Baxa, D. V., Lam, R., Miller, J. T., ... & Teh, F. C. (2016). Anthropogenic debris in seafood: Plastic debris and fibers from textiles in fish and bivalves sold for human consumption. *Scientific Reports*, 6(1), 1–10. https://doi.org/10.1038/srep26187.

Shim, W. J., Hong, S. H., & Eo, S. (2018). Marine microplastics: Abundance, distribution, and composition. In S. Galloway (Ed.), *Microplastic Contamination in Aquatic Environments* (pp. 1–26). Elsevier. https://doi.org/10.1016/B978-0-12-813747-5.00001-1.

Smith, M., Love, D. C., Rochman, C. M., & Neff, R. A. (2018). Microplastics in seafood and the implications for human health. *Current Environmental Health Reports*, *5*, 375–386.

Su, L., Xu, X., Li, G., & Wu, C. (2021). Microplastics in a shrimp farming system: Occurrence, composition, and potential pathways. *Environmental Pollution*, 278, 116870.

Sussarellu, R., Suquet, M., Thomas, Y., Lambert, C., Fabioux, C., Pernet, M. E. J., ... & Huvet, A. (2016). Oyster reproduction is affected by exposure to polystyrene microplastics. *Proceedings of the National Academy of Sciences*, 113(9), 2430–2435.

Wang, J., Peng, J., Tan, Z., Gao, Y., & Zhan, Z. (2018). Microplastics in aquatic environments and their potential effects on aquatic organisms. *Environmental Pollution*, 237, 494–504. https://doi.org/10.1016/j.envpol.2018.02.064.

Wang, Q., Wang, Y., Zhao, Y., Xu, Z., Zuo, Z., & Zhang, S. (2021). Chronic ingestion of microplastics impairs intestinal function and alters gut microbiota community in shrimp *Litopenaeus vannamei. Environmental Pollution*, 270, 116280. https://doi.org/10.1016/j.envpol.2020.116280.

Wright, S. L., Ulke, J., Font, A., Chan, K. L. A., & Kelly, F. J. (2020). Atmospheric microplastic deposition in an urban environment and an evaluation of transport. *Environment International*, 136, 105411.

Wu, J., Liu, Y., Wu, Y., Wang, H., & Li, X. (2021). Occurrence and potential human health risks of microplastics and associated contaminants in seafood from China. *Journal of Hazardous Materials*, 403, 123585. https://doi.org/10.1016/j.jhazmat.2020.123585.

Yang, Y., Zhang, X., Ma, Y., & Ma, F. (2021). Microplastics in fish feed: Current status and future perspectives. *Aquaculture*, 537, 736478. https://doi.org/10.1016/j.aquaculture.2021.736478.

Zhang, W., Wu, Y., & Jabeen, K. (2020). Microplastics in aquaculture: Transfer from feed to the gut of fish. *Environmental Pollution*, 264, 114771.

Zhou, A., Zhang, Y., Xie, S., Chen, Y., Li, X., Wang, J., & Zou, J. (2021). Microplastics and their potential effects on the aquaculture systems: A critical review. *Reviews in Aquaculture*, 13(1), 719–733.

Zhu, X., Chen, L., Liu, S., & Shi, H. (2020). Microplastics in aquatic organisms: A case study in China. *Frontiers in Environmental Science*, 8, 570596. https://doi.org/10.3389/fenvs.2020.570596.

7 Investigating Approaches and Technologies for Microplastics in Wastewater Management

Revantkumar S. Gorfad, Herish N. Ribadiya, Vinitkumar K. Rathod, Meet R. Parmar, and Suranjana V. Mayani

7.1 INTRODUCTION

The environment is adversely affected significantly and broadly by the unsustainable use and disposal of plastic. Microplastics are described as plastics with a diameter of less than 5 mm that are created on purpose or manufactured as larger plastics degrade. Environments and living things may be negatively impacted by microplastics (Van Cauwenberghe et al., 2015). Microparticles of plastics found in water pose a threat to public health because they can bind to harmful substances like drugs, hazardous bacterium (McCormick et al., 2014; Ziajahromi et al., 2017). Even wastewater is a significant contributor of microplastics, little research has been done on the topic. Primary microplastics are defined as plastic particles of the size that are specifically produced for use in products (such as toothpaste or exfoliants for cosmetics) or by businesses (such as air blasting) and which frequently come from improperly disposed of consumer goods. They slowly deteriorate due to photo- and thermo-oxidative processes and, to a lesser extent, biodegradation. This weakens the material's integrity and causes fragmentation into pieces smaller than 5 mm, or secondary microplastics (Andrady, 2011). Microplastics have already been identified in seawater at concentrations of up to 10 microplastic pieces per cubic meter. When mini-microplastics are taken into account, the recalculated number is more closely 8.3 million pieces per cubic meter (Andrady, 2011) and they have also been found to contaminate freshwater (Eriksen et al., 2014), sediment (Abidli et al., 2018), soil (Watteau et al., 2018), air (Dris et al., 2016; Abbasi et al., 2019) even food-related items like traditional salt, beer, and regular water (Kosuth et al., 2018). Human exposure to these particles cannot be prevented due to the widespread presence of environmental microplastics and microplastics in products for consumers.

The effects of this interaction, though, are not yet totally understood. This study conducts a thorough analysis of the information relating to the negative health impacts associated with environmental contact with plastic debris. In addition, hypotheses

DOI: 10.1201/9781003438793-7

for their toxicological mechanisms and exposure pathways are offered, possibly laying the groundwork for more. There is a scarcity of data on the consequences of microplastics on humans as a result of ethical constraints, strict biosecurity protocols for handling human samples, and constrained detection techniques. This means that discussions of risk and contamination also include data from testing on organisms.

Cosmetics may contain 0.5%–5% of primary microplastics, or microbeads, with a typical size of 250 μm. Microbeads have replaced natural exfoliants, particularly powdered walnut husks, in exfoliator cleansers, reducing skin irritation and damage (Chang, 2015). Microbeads in toothpaste work as an abrasive to remove plaque and stains (Vieira et al., 2016). In contrast to toothpaste's 4,000 microbeads, cleansing washes can produce between 4,500 and 94,500. The estimates of microbead release that are currently available only consider two categories of personal care products and one type of polymer (polyethylene), and they do not take into account actual retention rates in wastewater treatment plants (Carr et al., 2016; Napper et al., 2015). According to Boucher and Friot (2017), washing clothes produces fibers from synthetic textiles that make up to 35% of the microplastics found in the seas. According to many studies, a single garment may release up to 1,000,000 fibers, 110,000 (Almroth et al., 2017), and more than 1,900 (Browne et al., 2011). Polyester textiles normally release roughly 6,000,000 fibers when washed with equivalent weights (5–6 kg), whereas acrylic fabrics typically release 700,000 fibers (Napper and Thompson, 2016). According to Sillanpaa and Sainio (2017), washing machines in Finland emit between 154,000 and 411,000 kg of cotton microfibers and polyester per year (its length: 10,000–100,000 cm; thickness: 1,000–2,000 cm). The values shown are significantly influenced by the textile's composition (polymer), washing circumstances (friction, temperature washing duration, velocity), use and type of softener and detergent, and weathering of clothing, in addition to differences in research design.

Suggested by Napper et al. (2015) that spectacles cleaners, glitter, smaller-sized buttons, bottles, ornaments, and different commercial goods may also introduce microplastic particles into waste water systems. Microplastics in wastewater could come from a variety of non-domestic sources, including: lost pre-production pellets that were consumed during transportation or manufacturing; air blasting using plastic particles to clean engines of paint and grime; synthetic fibers used in the textile industry; plastic dust which is foam when cutting the plastics; lost Styrofoam used in fillings or shipping; and dust from drilling and cutting plastics. When these particles are lost, they could by mistake get into effluent or draw down systems. It is critical to examine these sources' contributions further while taking into consideration polymer types, whole different products and generating strategies for limiting their losses.

Microplastics are consumed by aquatic creatures, carried up the food chain by water bodies, and ultimately cause irreparable harm to human health (Pironti et al., 2021). Home, business, and agricultural wastewater sources all provide considerable amounts of microplastic-filled wastewater to wastewater treatment plants on a daily basis (Liu et al., 2021b). Numerous researches have tried to understand how plastic particles behave during the treatment process with different degrees of success based on the kind of wastewater and the treatment method employed. In the main clarifier, testing material being removed and settling procedures just physically remove large and low-density microplastics, whereas flotation removes large-shaped microplastics like fibers and fragments. Gravity usually causes high-density microbeads to fall to the bottom of the sedimentation tank. Huge debris particles and huge microplastics are both targeted for removal during the physical treatment method in order to maintain the facility's performance for the subsequent treatment operation (Lares et al., 2018; Liu et al., 2021a, 2021b). Furthermore, microscopic microplastic particles and microorganisms are mixed to create flakes, which are subsequently fed to the extra clarifier for biological treatment (Ziajahromi et al., 2017; Sun et al., 2019). A number of processing processes and cutting-edge technologies, such as rapid sand filters, disc filters, air flotation or ozonation processes may be used to remove 90%–98% of the microplastics in wastewater, according to a prior study (Mintenig et al., 2017). The bulk of earlier studies (Liu et al., 2019; Mahon et al., 2017) primarily concentrated on the quantity of microplastics in the effluent and came to the conclusion that WWTPs were a source of plastic particles in the bodies of water. Activated sludge made from recycled plastics poses an extra risk from microplastics. Sludge is frequently used in agriculture as fertilizer or as concrete and eventually causes environmental degradation on land (Magni et al., 2019).

In light of the growing concern over microplastic pollution in aquatic environments, this study investigates a comprehensive analysis of various approaches and technologies aimed at effective microplastic management within wastewater systems. By investigating the strengths and limitations of existing methodologies, this research strives to provide valuable insights that can guide the development of innovative and sustainable solutions. Through rigorous examination and comparison, we aim to contribute to the ongoing efforts to mitigate the presence of microplastics in wastewater and ultimately safeguard the health of our ecosystems and the well-being of future generations.

7.2 DIFFERENT TYPES OF MICROPLASTICS IN WASTEWATER

Microplastics are first divided into two parts, primary and secondary groups, according to production (Cole et al., 2011; Laskar and Kumar, 2019). Primary microplastics are purposefully produced because of their commercial value. They serve a variety of purposes in society, including the manufacture of small microbeads for use in beauty products, personal care items, cosmetics, detergents, medical apparatus, pharmaceuticals items, different types of inks and paints, concrete, polymer cement, coatings, some type of paper, sewage sludge dewatering, polishing agents, horticulture, and agriculture are some of the related technologies (Fan et al., 2021) Secondary MPs are fragments left over after larger-sized plastic products including bags, bottles, clothing, pipelines, nets, covers, doors, ropes, fan blades, and tires. decay into smaller-sized plastic particles (Reisser et al., 2013). Plastic degradation occurs through biotic and abiotic mechanisms, with biotic degradation involving microorganisms breaking down biodegradable materials, while abiotic degradation involves thermal, photocatalytic, and mechanical forces breaking down larger materials (Sudharshan Reddy and Abhilash Nair, 2022).

Due to household discharges (such as washing clothes), rain runoff, and street washing, all type of MPs frequently enters into different source of water in a significant manner. MPs have physiochemical characteristics that vary from plant to plant, including different sizes, shapes, colors, and types of polymers. Several samples of untreated wastewater contained particles, fibers, films, microbeads, foam, granules, pellets, and spherical and non-spherical forms. However, fibers and pieces contributed the majority of the form (Sudharshan Reddy and Abhilash Nair, 2022). Plastic pellets are granular materials with a typical diameter of 2–5 μm and a uniform shape, which are employed in the manufacture of various plastic products. Typically, semi-finished goods are produced, shipped, and stored as plastic pellets. Polyethylene, propylene, polystyrene, polyvinyl chloride, and other ingredients used to make plastic are derived from petroleum and coal. Plastic fishing waste includes floating boxes, fishing poles, nets, lines, and cables, fish tanks, among other items. On a global scale, it is estimated that between 0.13 and 135,000 tonnes of commercially used fishing equipment are abandoned each year.

7.2.1 Microbeads as a Microplastic

Microbeads are any synthetic, "non-biodegradable," solid microplastic particles that are intentionally added to various goods and operations and maintain their predetermined shapes "during life cycle and after disposal." Microbeads (MBs) are the microplastic particles with a size range of approximately 5 μm to 1 μm that are primarily used as to made personal care products. This criterion allows many MBs that are currently on the market, even if they only partially biodegrade in aquatic environments. Since microbeads (MBs) were invented for use in cosmetics in the 1970s, the usage of activating cleansers containing plastic ingredients has dramatically expanded. Size, shape, and polymers' composition, MBs, which are often sold, differ according to the product. Cosmetics manufacturing uses the phrase "microbead" to denote any rough plastic particle, despite the fact that it may conjure up thoughts of small, spherical beads in vivid colors. Hand/face cleaners, toothpaste, peels, lotions, lipsticks, mascara, blushers, hair color, baby products, nail polish, scrubs, sunscreens, shower gels, and shaving creams are widely used examples of cosmetics that include synthetic plastic microbeads (MBs). These products are widely used (Yurtsever, 2019). Figure 7.1 shows the various types of microplastic reported by earlier researchers.

7.2.2 Microfibers as a Microplastic

Microplastics are traditionally defined as polymer particles or fragments with a diameter of less than 5 μm (Acharya et al., 2021). Microplastic fibers can be produced even before large plastic particles in the environment, such as when washing clothes, or they can be broken down into smaller pieces in the environment. Microplastic fibers are further characterized in two groups: filament and staple fiber, based on the ratio of fiber length to diameter. With the exception of silk, all types of natural fibers and synthetic filaments can be cut into staples to create fibers. Staple fibers typically have lengths between 3 and 20 cm. Fibers of the type cotton (2–3 cm) and wool (5 cm or more) are referred to as "short staple" and "long staple," respectively. According to Liu et al. (2019), among the many different types of microplastics found in the environment, the fibrous/filamentous form predominates in both marine and freshwater ecosystems. There is good evidence to believe that synthetic fabrics, particularly polyester,

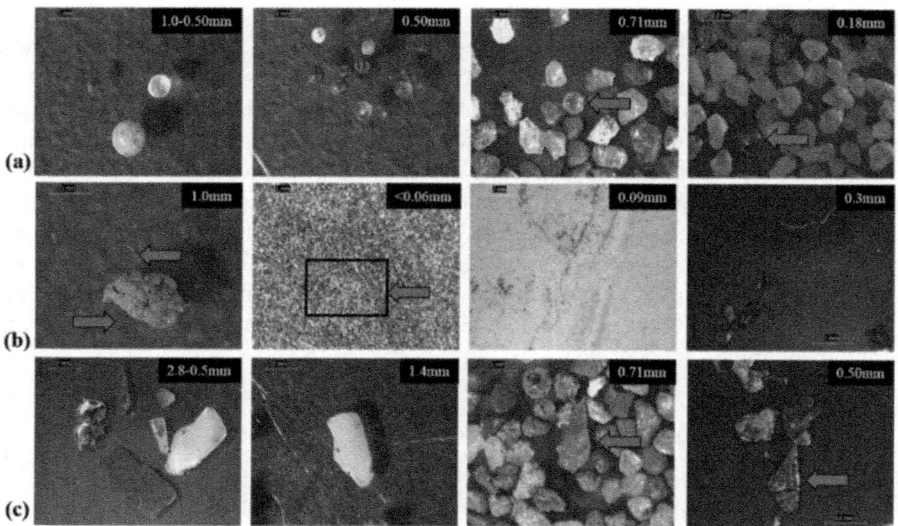

FIGURE 7.1 Light microscopy images of suspected microplastics in size-fractionated sediment samples from the River Kelvin in suspended and settled material before chemical characterization. Items shown are pellets (a), fibers (b), and fragments (c). (Reina M. Blair et al., 2019).

are a significant source of microplastic fibers given the continued growth in their use (Hernandez et al., 2017). The majority of the different types of plastic particles were fibers and lines. The majority of the plastic particles classified as fibers/lines were fibers rather than lines. Reported that human-made microfibers were present throughout the river. Acharya et al. (2021) reported that 100 million tonnes of fibers, including both natural and synthetic fibers, were produced worldwide in 2017. Man-made fibers include synthetic and artificial fibers. The world's consumption of fibers is currently dominated by synthetic materials, with polyester, polyamide, acrylic, and polyolefin being the most widely used types. However, only artificial textiles can be categorized as plastic. The fact that not all of the fibers in the environment are made of plastic, however, is a fact (Xu et al., 2018; Šaravanja et al., 2022).

7.2.3 Glitter as a Microplastic

Several different types of small, fatty, shiny particles are referred to as "glitter." Glitter is manufactured from the plastic polymer MylarTM, also referred as BoPET (biaxially oriented polyethylene terephthalate). The polyester, also said to be BoPET, is manufactured from stretched polyethylene terephthalate (PET). The glitter (PET) particles' melting point is 260°C and a real density of 1.38 g/cm^3. They are not soluble in water. The particles have layers of metal (aluminum-coated polyethylene terephthalate glitter) to achieve great reflectivity. Plastic glitters are available in silver and golden colors. They can even make themselves appear holographic. Glitters are made in a variety of shapes (in consistently sized, carefully cut pieces, with notches on occasion): the most popular shapes are hexagonal, and they include, triangle, square, octagonal, stripe, heart, star, moon, diamond/rhombus, flower, snowflake, butterfly, and also different irregular shapes. Producers routinely promise to supply 10,000 different combinations of glitter colors, shapes, and sizes on e-commerce sites. The most prevalent and adaptable size of polyester glitter is 200 µm (0.008″), which is the typical size for imitation glitter. Glitters are microplastics since they are created in large quantities using PET fibers (polyester) and are sold in sizes less than 5 µm (Yurtsever and Yurtsever, 2018).

7.3 METHODS AND APPROACHES

7.3.1 Occurrence and Sources

In WWTP influent and effluent, 30 different types of micro-level plastic polymers have been detected. Most polymers seen in the effluents and powerful of the WWTPs were polyester (28%–89%), polyethylene either polythene (4%–51%), polyethylene terephthalate (4%–35%), and polyamide (3%–30%) (Mintenig et al., 2017; Ziajahromi et al., 2017; Cheung and Fok, 2017; Lares et al., 2018) PET, polyethylene terephthalate (PES), and PA. The effluent also contained polymers with concentrations ranging from 5% to 27%, including styrene, alkyd, polyurethane, propylene, vinyl alcohol, acrylate, and lactide. The total amount of microplastics found in sewage was typically less than 5% and could possibly be as low as 1% in some samples, consisting of only a small portion of the remaining polymers. As a result, the most common polymers might be given priority for study rather than all of the offered particles.

7.3.2 Treatment Methods

The WWTP's capacity to remove microplastics was evaluated using the microplastic concentrations in the wastewater and waste product. With the exception of the study mentioned (Leslie et al., 2017), the overall effectiveness of eliminating microplastics from wastewater was higher than 88% in WWTPs without tertiary treatment and higher than 97% in WWTPs with tertiary treatment. Relatively poor removal effectiveness may be explained by the fact that they only collected 2L of wastewater for the purpose of identifying microplastics, compared to the tens to thousands of liters collected for analysis in the majority of other studies. However, as noted in this study, it could also be brought on by the failing efficiency of some reactor types, like membrane reactors (Leslie et al., 2017).

7.3.2.1 Preliminary Treatment and Pre-treatment

Initial treatment may be used to remove most of the MPs detected in wastewater. It is anticipated that about 35%–59% of MPs can be removed with the help of preliminary treatment, and 50%–98% can be removed after the primary therapy. During grit removal and gravity separation, heavy MPs or microplastics locked in solid flocs were generally settled, and during grease skimming or surface skimming, light suspended microplastics were scanned on primary clarifiers.

The pre-treatment successfully removed larger-size microplastics, which had the biggest effect on the microplastic size delivery. Dris et al. (2015) detected that the percentage of big particles removed from 45% to 7% after the first treatment. Taking into account the proportionate number of fibers that are decreased after pre-treatment, research on the morphologies of microplastics suggests that pre-treatment may be more successful than fragmentation at removing fibers from wastewater (Magnusson and Norén, 2014; Talvitie et al., 2015; Ziajahromi et al., 2017). This is most likely because fibers can be separated by sedimentation and more readily captured by flocculating particles. A significant number of these tiny fragments are formed out of PE, which is very floating in water and was probably going to float on top of the wastewater or grease, particles of oil and fat where they could be skimmed off rapidly, as noticed (Murphy et al., 2016). This result was strengthened by the outcomes of two research (Sutton et al., 2016), both of which discovered that small beads were missing from WWTP effluent. Four out of ten of the WWTPs in New York, USA, continue to discharge microbeads, according to a research study (Schneiderman, 2015). Different amounts of grease, fat, and particle of oil in the sewage could be attributed to this variation because these substances may aid in the removal of microplastics from the surface of the water.

7.3.2.2 Secondary Treatment

Plastic particles in wastewater were decreased to 0.2%–14% with secondary treatment (typically biological action and clarity). At this step, it's likely that bacterial extracellular polymers or sludge flocs that form in the aeration tank and are subsequently deposited in the secondary clarifying tank will help capture any remaining plastic debris. A "floc" is formed when particles in suspension cluster together as a result of chemicals used in secondary treatment, such as ferric sulfate or other flocculating agents (Murphy et al., 2016). It was unclear, however, how plastic particles reacted with microorganisms or chemical-based flocs or how much this might aid in the clearance of microplastics. A few microplastics may also get caught up in unstable flocs and struggle to settle, which causes them to move around in the aqueous phase and avoid being removed throughout the subsiding stage (Carr et al., 2016).

For the purpose of removing microplastics from secondary discharges, it is thought that the length periods that microplastics interact with the pollutants in the treatment train is crucial. In 2016, Carr et al. (2016) found that the inclination for the microplastics' bacterial-coated surface increased with duration of contact. As wetting agents, these bio-coatings may change the surface traits or relative densities of the microplastics. Rummel et al. (2017) and subsidiaries. Neutrally floating particles are more likely to avoid both skim and settling techniques. Hence, the success rate of removing microplastics might be strongly influenced by such changes. More studies should be done on the impact of contact time and the amount of nutrients on the surface clogging and removal effectiveness of MPs. Later studies in wastewater systems will be well-supported by the investigational techniques and scientific representations used for the study of the development of biofilm on MPs and the impact on particle transport in freshwater and aquatic environments (Besseling et al., 2017; Rummel et al., 2017; Fazey and Ryan, 2016).

The secondary treatment process eliminated supplementary component particles from fibers in comparison to the pre-treatment. The proportionate prevalence of microplastic pieces was reduced, but that of fibers improved after the subsequent dealing, according to studies (Talvitie et al., 2015, 2016; Ziajahromi et al., 2017). One possibility is that during pre-treatment, the readily settling or skimming fibers were already greatly eliminated, but the remaining fibers may have possessed characteristics like neutral buoyancy that made them difficult to remove further.

As a result of further removal of these particles during secondary treatment, a relatively low abundance of big microplastics can be achieved in the secondary effluent. Microplastics greater than 500 μm in size were discovered to be basically non-existent in secondary waste (Mintenig et al., 2017; Talvitie et al., 2016), with microparticles larger than 300 m accounting for only 8% of the total after the secondary treatment process. In disparity, after secondary treatment, Dris et al. (2015) found that MPs with a size between 500 and 1,000 m still made up 43% of the total. It was unclear why there was such a huge percentage. It may be correlated to the distinct MPs elimination efficacy reached through several secondary treatment procedures under various operative situations, which requires further investigation in the future.

7.3.2.3 Advanced Treatment

An advanced treatment technology could provide significant solutions for the removal of MPs. The global number of plastic particles which is present in the wastewater was reduced to 0.2% of the influent after tertiary treatment process. The treatment processes used influence the success of microplastic removal, with membrane-related technologies doing well. Talvitie et al. (2017) investigated the efficacy of numerous tertiary treatments for secondary effluent, counting dissolved air flotation (DAF), disc filter (DF), membrane bioreactor (MBR), and rapid sand filtration (RSF) for primary effluent. They discovered that dissolved air flotation and membrane bioreactor had the maximum removal efficacy (99.9%), followed by rapid sand filtration and dissolve air flotation, which had 97% and 95% removal rates, correspondingly. The elimination efficacy for Discfilter is ranged from 40% to 98.5%. Likewise, in the previously mentioned assessment of WWTPs in New York (US), two membrane-filtering shrubberies did not produce microbeads, although the further four plants with more progressive filters fixed (i.e., a Fast Sand Filter, Constant Backwash Filter, and two filters of unknown kind) (Schneiderman, 2015). Furthermore, Ziajahromi et al. (2017) detected that MP concentrations were considerably abridged after reverse osmosis and ultrafiltration. According to Mason et al.'s (2016) research, granular-improved filtration did not lower the quantity of MPs emitted by WWTP.

The effluent and influent of the third-treatment unit may include a small amount of microplastics (1 particle/L in the majority of situations), which is an important fact to keep in mind, which can cause misleading zero findings to be obtained from small sample volumes (dozens of liters). As a result, larger sampling volumes are required to appropriately compare tertiary treatment methods to pre-treatment and secondary treatment procedures in terms of their ability to remove microplastics. During tertiary therapy, it was found that the smallest sizes (20–100 m and 100–190 m) were the most common (Ziajahromi et al., 2017). This is likely because fibers facilitate the longitudinal passage of any filter or barrier. As a result, the necessity of final-stage technology for eliminating particularly tiny and the emphasis is on fiber-like MPs from effluent. Normally, solids gathered during filter backwashing are taken back to the WWTP's origin. As a result, microplastics from the third treatment are likely not eliminated through the WWTP and might boost their microplastic loading. This fraction of the microplastics can be reduced by pre-treatment or additional treatment as the contact time improves. However, the chance of their escaping from the treatment processes may increase if microplastics are involved in the effluent.

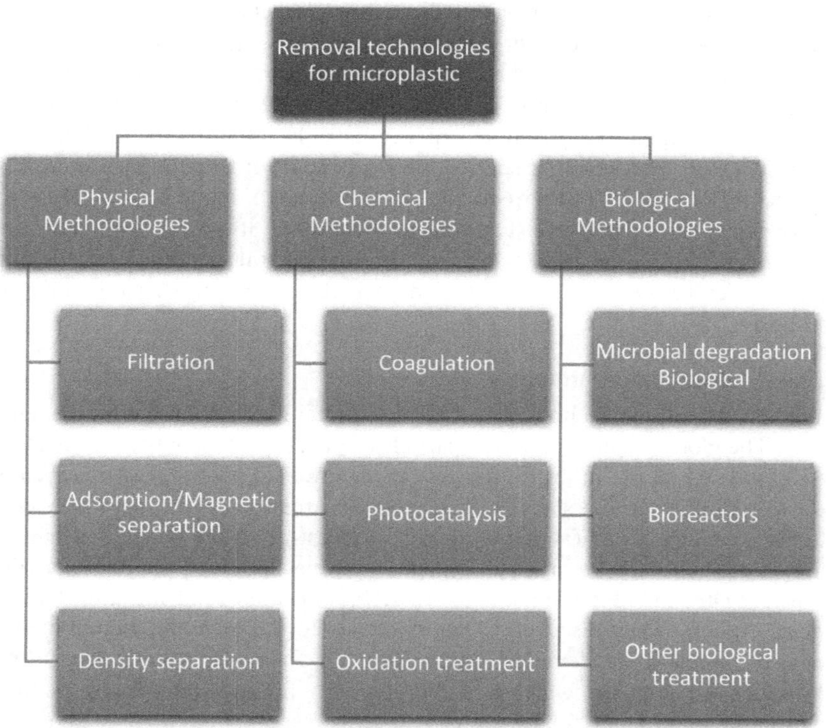

FIGURE 7.2 Overview of technologies for removal of MPs in water.

7.4 REMOVAL TREATMENT TECHNIQUES

Figure 7.2 provides an overview of technologies that have been potentially employed for microplastic removal from an aqueous environment.

7.4.1 Physical Methodologies

7.4.1.1 By Filtration

The most popular and appropriate method for removing minute particles from water is filtration. To remove MPs from the aquatic environment, a variety of filtration procedures have been attempted (Ali et al., 2021). The fact that 180 nm particles were removed more readily than 1–20 m particles suggested that small particles were more easily retained through particle attachment or diffusion. However, the effluent contained particles that were 10–20 m in size. In biochar filters, MPs were primarily immobilized by "stuck," "trapped," and "entangled" processes, whereas the MP spheres contained just "Stuck" in sand filters. In both WWTP and DWTP, membrane filter technology has been extensively utilized. Several membrane filtration techniques, including ultrafiltration, microfiltration, membrane bioreactors, reverse osmosis, and dynamic membrane, have been utilized to reduce MP contamination.

Additionally, DWTP utilized media filtering techniques, such as sand filtration and the application of activated carbon particles (Figure 7.3a) (Gao et al., 2022). Microplastics have been effectively eradicated from polluted marine environments by employing membrane technologies. The number, size, flux, and durability of the microplastics in the membrane all have a significant impact on how well the particles are removed. Combining biological processes and porous membranes could raise the removal effectiveness to 99.9% (Padervand et al., 2020).

7.4.1.2 By Adsorption

Adsorbents and MPs interact during the adsorption process, which is the basis for the adsorption method. Adsorption technique elimination has a very good efficacy in eliminating nanoplastics (NPs) and small microplastics (MPs), particularly for particles less than 10m. The adsorption mechanism incorporates interactions such as H-bonds, electrostatic forces, and contacts (Figure 7.3b). Several academics employed crushed or small-scale adsorbents to remove MPs. As an example, Tiwari et al. (2020) Synthesized Zn-Al layered double hydroxide (LDH) through co-precipitation, along with subsequent testing, demonstrated that LDH's highest capacity for adsorption of 164.49 mg/g for 55 nm PS MPs in deionized water, with a 96% removal rate of all MPs, proved possible. After mixing and aerating particulate adsorbents and PS in the liquid phase to remove adsorption, the resultant mixtures were centrifuged to separate them to generate a clean supernatant (Tiwari et al., 2020; Gao et al., 2022).

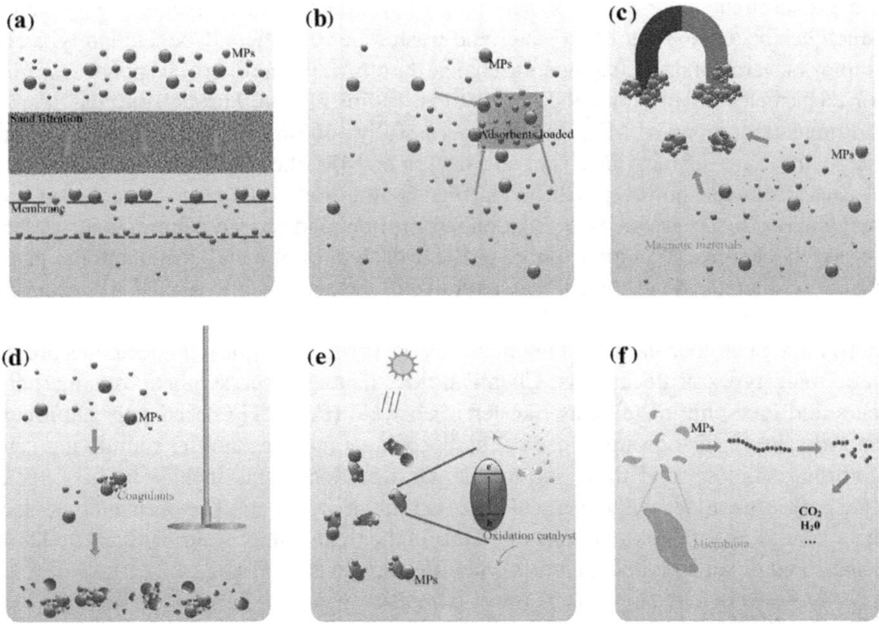

FIGURE 7.3 Schematic diagrams of main technologies for removing MPs in water. Filtration technology (a), adsorption removal (b), magnetic removal (c), coagulation treatment (d), photocatalysis (e), and biodegradation treatment (f) (Gao et al., 2022).

7.4.1.3 Magnetic Removal

Magnetic separation has recently been developed for MP removal; wide-surface area magnetic nanoparticles were combined with MPs in the removal process using adsorbents. MPs may be quickly and easily retrieved from water using magnetic force after being magnetized (Figure 7.3c). The magnetization of MPs is a result of a variety of mechanisms. The great hydrophobicity of MPs contributed significantly to the magnetic carbon nanotube adsorption (M-CNTs) by PE, and interactions between electrons and hydrogen atoms through complexation, electrostatic interaction, and H-bonds resulted in the adsorption of M-CNTs by polyamide (PA). (Gao et al., 2022; Tang et al., 2020). The size and kind of MPs, water pH, magnetic particle concentrations, and other variables will all have an impact on how effectively they are removed. According to a recent study, MPs can be removed by magnet adsorption and magnetization using nano-Fe_3O_4 The four most common MP kinds' mean removal rate, PE, PP, PS, and PET, ranged from 62.83% to 86.87% (Shi et al., 2022). Another study showed that PA MPs with positive charges increased their opposition to M-CNTs via electrostatic forces in acidic surroundings, however, the formation of acidic oxygen-containing groups on the carbon nanotube surface diminished the hydrophobic effect of PE/PET and M-CNTs (Tang et al., 2020).

7.4.2 Chemical Methodologies

7.4.2.1 Coagulation

The predominant method for pollutant removal in water treatment plants is coagulation, which has been widely used in water and wastewater treatment. Coagulation typically employs several kinds of coagulants to change the physical condition of suspended and dissolved particles and promotes their removal via sedimentation. Through the coagulation/sedimentation process, MPs can be successfully eliminated. Even though various coagulants have distinct impacts of elimination on MPs, the individual removal process of each coagulant however will be clarified by the traditional removal techniques of coagulation, such as sweep flocculation, adsorption, and charge neutralization. When negatively charged, MPs are exposed to the hydrolysates of metal coagulant, the initial charge on the outside of the MPs is neutralized, the electrostatic repulsion is reduced, and the MPs become unbalanced (Xu et al., 2021; Zhou et al., 2020). Chemical flocculants, biological flocculants that are naturally occurring, and grafted flocculants are the three main types of flocculants. Chemical flocculants include synthetic organic polymers and inorganic metal salts like ferric chloride ($FeCl_36H_2O$) and poly aluminum chloride (PAC). The group of natural bioflocculants includes tannin, sodium alginate, chitosan, cellulose, and these substances. Grafted flocculants include both synthetic (like polyacrylamide) and natural (like grafted starch) polymers. The dormant to replace the potential use of inorganic salts of metals in the treatment of water with grafted flocculants and organic bioflocculants is great (Peydayesh et al., 2021).

7.4.2.2 Photodegradation

It is believed that photodegradation is a sophisticated form of oxidation that uses active species like hole (H^+), O_2, and OH produced through photocatalysts when exposed to light to degrade contaminants into water and CO_2. (Zhou et al., 2020).

To photodegrade various kinds of MPs, various photocatalytic materials have been created. The deterioration of MPs has drawn a lot of attention to TiO_2, a traditional photocatalyst (Ariza-Tarazona et al., 2018). First, it was reported that two photocatalysts based on $NTiO_2$—protein-resultant $NTiO_2$ and solgel $NTiO_2$—were used to photo-catalyze the breakdown of high-density polyethylene microplastics. The effectiveness of two photocatalysts was assessed in this investigation during simulated irradiation with visible light protein-derived $NTiO_2$ shown improved polyethylene weight loss MPs promotion than sol–gel $NTiO_2$ (6.40% vs. 2.86%). Polystyrene MPs had higher degradation ability than reported utilizing TiO_2 under simulated ultraviolet light irradiation compared to minimal loss of weight of MPs made of polyethylene (Wang et al., 2019). By avoiding the negative effects of the polystyrene MPs hydrophilic properties in the solution of water and the reaction, it was discovered, for instance, that after 12 hours, the disintegration efficiency of 5 mm polystyrene MPs reached 44.66% of illumination under 254 nm ultraviolet. This value improved to 99.99% in the solid phase under identical circumstances. The bigger polystyrene MP fragments were divided into smaller particles throughout this process, and the predominant byproduct was CO_2 (Nabi et al., 2020).

7.4.3 Biological Methodologies

7.4.3.1 By Microorganism

This process is carried out by microorganisms with the aid of enzymes found in living things that are most able to break down such plastics from various inorganic components, biomass, CH_3, CO_2, and water. For microplastic biological degradation, polymer plastic types with certain chemical and physical characteristics are needed. Among the environmental elements that influence biotic degradation are natural light, UV radiation, temperature, and air humidity (Dey et al., 2021).

7.4.3.2 Fungal Degradation

Polymers can serve as food for a variety of fungi (Russell et al., 2011). Consequently, such microbes could be used to get rid of MPs. The aquatic fungi (*Zalerion maritimum*), which are broadly dispersed in the marine waters of Portugal, have been utilized by the researchers to degrade PE-based MPs. After exposing the fungus to MPs for 14 days, under restricted conditions (25°C, darkness circumstances, 120-rotation per minute constant swirling), it was observed that there was a positive link between a rise in terms of percentage of fungal mass and a reduction in the percentage of plastic weight. The fluctuation in biomass in this case was 82%, the variation of plastic weight was 56.7%, and the elimination was greater than 43% (Paço et al., 2017). In 2011, research was also carried out to see if endophytic fungi (*Pestalotiopsis microspora*) would be degrading Polyester Polyurethane (PUR) MPs. Although PUR is thought of as the source of carbon, serine hydrolase was revealed to be the catalyst for this polymer's degradation in an anaerobic environment (25°C) (Russell et al., 2011).

7.4.3.3 By Algae

Sundbæk et al. (2018) analyzed how the edible marine microalga fucus vesiculosus was covered in luminous microplastic particles. To stop the polystyrene microplastics from entering the tissues, the sorbent's plant cells had extraordinarily small microchannels. The microplastics had a diameter of 20 m. The function of cell wall-released alginate molecules in the areas of cuts can be used to explain the results, which showed a significant surface assimilation of MPs (94.5%), mainly adjacent to the seaweed's sliced edges. The sorption of microplastics on algal surfaces is generally significantly influenced by the surface charge of the microbes. An negatively charged polysaccharide is present in the chemical composition of the algal cell, which indicates that the cationic in nature microplastics have a higher tendency to be sorbed more successfully (Sundbæk et al., 2018).

7.5 CONCLUSION AND FUTURE PERSPECTIVES

In conclusion, the research into methods and technology for microplastics in wastewater treatment has highlighted how urgent it is to deal with this widespread environmental issue. Microplastics in wastewater pose serious threats to aquatic ecosystems as well as to human health. It is clear from a thorough analysis of numerous tactics that a multifaceted strategy is necessary for successful microplastic abatement. The study emphasized the significance of source control measures as a first line of defense against microplastics entering wastewater systems, including a decrease in plastic usage and enhanced waste management techniques. Additionally, cutting-edge methods for wastewater treatment such as membrane filtration, coagulation, and activated carbon adsorption present intriguing options for catching and getting rid of microplastics from wastewater streams. However, it is crucial to understand that due to variances in wastewater composition, treatment facilities, and regional variables, a one-size-fits-all solution may not be appropriate. In order to unify regulatory frameworks, encourage innovation, and adopt sustainable practices, cooperation between governments, industries, researchers, and communities is essential. In order to comprehend the long-term impacts of microplastics on ecosystems and human health, more research is required, as the investigation has highlighted. Furthermore, ongoing technical development and interdisciplinary research will outgrowth the creation of more effective, affordable, and eco-friendly microplastic cleanup techniques.

REFERENCES

Abbasi, S., Keshavarzi, B., Moore, F., Turner, A., Kelly, F.J., Dominguez, A.O., Jaafarzadeh, N. (2019). Distribution and potential health impacts of microplastics and microrubbers in air and street dusts from Asaluyeh County, Iran. *Environmental Pollution* 244, 153–164. https://doi.org/10.1016/j.envpol.2018.10.039.

Abidli, S., Antunes, J.C., Ferreira, J.L., Lahbib, Y., Sobral, P., Menif, N.T.E. (2018). Microplastics in sediments from the littoral zone of the north Tunisian coast (Mediterranean Sea). *Estuarine, Coastal and Shelf Science* 205, 1–9. https://doi.org/10.1016/j.ecss.2018.03.006.

Acharya, S., Rumi, S.S., Hu, Y., Abidi, N. (2021). Microfibers from synthetic textiles as a major source of microplastics in the environment: A review. *Textile Research Journal* 91(17–18), 2136–2156. https://doi.org/10.1177/0040517521991244.

Ali, I., Ding, T., Peng, C., Naz, I., Sun, H., Li, J., Liu, J. (2021). Micro- and nanoplastics in wastewater treatment plants: Occurrence, removal, fate, impacts and remediation technologies – A critical review. *Chemical Engineering Journal* 423, 130205. https://doi.org/10.1016/j.cej.2021.130205.

Almroth, B.M.C., Astrom, L., Roslund, S., Petersson, H., Johansson, M., Persson, N.-K. (2017). Quantifying shedding of synthetic fibers from textiles: A source of microplastics released into the environment. *Environmental Science and Pollution Research*. https://doi.org/10.1007/s11356-017-0528-7.

Andrady, A.L. (2011). Microplastics in the marine environment. *Marine Pollution Bulletin* 62(8), 1596–1605. https://doi.org/10.1016/j.marpolbul.2011.05.030.

Ariza-Tarazona, M.C., Villarreal-Chiu, J.F., Barbieri, V., Siligardi, C., Cedillo-González, E.I. (2018). New strategy for microplastic degradation: Green photocatalysis using a protein-based porous N-TiO_2 semiconductor. *Ceramics International*. https://doi.org/10.1016/j.ceramint.2018.10.2.

Besseling, E., Quik, J.T.K., Sun, M., Koelmans, A.A. (2017). Fate of nano- and microplastic in freshwater systems: A modeling study. *Environmental Pollution* 220, 540–548.

Blair, R.M., Waldron, S., Phoenix, V.R., Gauchotte-Lindsay, C. (2019). Microscopy and elemental analysis characterisation of microplastics in sediment of a freshwater urban river in Scotland, UK. *Environmental Science and Pollution Research* 26(12), 12491–12504. https://doi.org/10.1007/s11356-019-04678-1.

Boucher, J., Friot, D. (2017). *Primary Microplastics in the Oceans: A Global Evaluation of Sources*. IUCN, Gland, Switzerland. https://doi.org/10.2305/IUCN.CH.2017.01. (43 pp.).

Browne, M.A., Crump, P., Niven, S.J., Teuten, E., Tonkin, A., Galloway, T., Thompson, R. (2011). Accumulation of microplastics on shorelines worldwide: Sources and sinks. *Environmental Science & Technology* 45, 9175–9179.

Carr, S.A., Liu, J., Tesoro, A.G. (2016). Transport and fate of microplastic particles in wastewater treatment plants. *Water Research* 91, 174–182. https://doi.org/10.1021/es201811s.

Chang, M. (2015). Reducing microplastics from facial exfoliating cleansers in wastewater through treatment versus consumer product decisions. *Marine Pollution Bulletin* 101, 330–333. https://doi.org/10.1016/j.marpolbul.2015.10.074.

Cheung, P.K., Fok, L. (2017). Characterisation of plastic microbeads in facial scrubs and their estimated emissions in Mainland China. *Water Research* 122, 53–61.

Cole, M., Lindeque, P., Halsband, C., Galloway, T.S. (2011). Microplastics as contaminants in the marine environment: A review. *Marine Pollution Bulletin* 62, 2588–2597. http://doi.org/10.1016/j.marpolbul.2011.09.025.

Dey, T.K., Uddin, M.E., Jamal, M. (2021). Detection and removal of microplastics in wastewater: Evolution and impact. *Environmental Science and Pollution Research* 28, 16925–16947.

Dris, R., Gasperi, J., Rocher, V., Saad, M., Renault, N., Tassin, B. (2015). Microplastic contamination in an urban area: A case study in Greater Paris. *Environmental Chemistry* 12(5). https://doi.org/10.1071/EN14167.

Dris, R., Gasperi, J., Saad, M., Mirande, C., Tassin, B. (2016). Synthetic fibers in atmospheric fallout: A source of microplastics in the environment. *Marine Pollution Bulletin* 104(1–2), 290–293. https://doi.org/10.1016/j.marpolbul.2016.01.006.

Eriksen, M., Lebreton, L.C.M., Carson, H.S., Thiel, M., Moore, C.J., Borerro, J.C., Galgani, F., Ryan, P.G., Reisser, J. (2014). Plastic pollution in the world's ocean: More than 5 trillion plastic pieces weighting over 250,000 tons afloat at sea. *PLoS ONE* 9(12), e111913. https://doi.org/10.1371/journal.pone.0111913.

Fan, C., Huang, Y.-Z., Lin, J.-N., Li, J. (2021). Microplastic constituent identification from admixtures by Fourier-transform infrared (FTIR) spectroscopy: The use of polyethylene terephthalate (PET), polyethylene (PE), polypropylene (PP), polyvinyl chloride (PVC) and nylon (NY) as the model constituents. *Environmental Technology & Innovation* 23, 101798. https://doi.org/10.1016/j.eti.2021.101798.

Fazey, F.M.C., Ryan, P.G. (2016). Biofouling on buoyant marine plastics: An experimental study into the effect of size on surface longevity. *Environmental Pollution* 210, 354–360.

Gao, W., Zhang, Y., Mo, A., Jiang, J., Liu, Y., Cao, X., He, D. (2022). Removal of microplastics in water: Technology progress and green strategies. *Green Analytical Chemistry* 3, 100042. https://doi.org/10.1016/j.greeac.2022.100042.

Hernandez, E., Nowack, B., Mitrano, D.M. (2017). Polyester textiles as a source of microplastics from households: A mechanistic study to understand microfiber release during washing. *Environmental Science & Technology* 51(12), 7036–7046. https://doi.org/10.1021/acs.est.7b01750.

Kosuth, M., Mason S.A., Wattenberg E.V. (2018). Anthropogenic contamination of tap water, beer, and sea salt. *PLoS ONE* 13(4), e0194970. https://doi.org/10.1371/journal.pone.0194970.

Lares, M., Ncibi, M.C., Sillanpää, M., Sillanpää, M. (2018). Occurrence, identification and removal of microplastic particles and fibers in conventional activated sludge process and advanced MBR technology. *Water Research* 133, 36–246.

Laskar, N., Kumar, U. (2019). Plastics and microplastics: A threat to environment. *Environmental Technology & Innovation* 14, 100352. https://doi.org/10.1016/j.eti.%20 2019.100352.

Leslie, H.A., Brandsma, S.H., van Velzen, M.J., Vethaak, A.D. (2017). Microplastics en route: Field measurements in the Dutch river delta and Amsterdam canals, wastewater treatment plants, North Sea sediments and biota. *Environment International* 101, 133.

Liu, J., Yang, Y., Ding, J., Zhu, B., Gao, W. (2019). Microfibers: A preliminary discussion on their definition and sources. *Environmental Science and Pollution Research* 26, 29497–29501. https://doi.org/10.1007/s11356-019-06265-w.

Liu, W., Zhang, J., Liu, H., Guo, X., Zhang, X., Yao, X., Cao, Z., Zhang, T. (2021a). A review of the removal of microplastics in global wastewater treatment plants: Characteristics and mechanisms. *Environment International* 146, 106277.

Liu, X., Yuan, W., Di, M., Li, Z., Wang, J. (2021b). Transfer and fate of microplastics during the conventional activated sludge process in one wastewater treatment plant of China. *Chemical Engineering Journal* 362, 176–182.

Magni, S., Binelli, A., Pittura, L., Giacomo, C., Della, C., Carla, C., Regoli, F. (2019). The fate of microplastics in an Italian wastewater treatment plant. *Science of the Total Environment* 652, 602–610.

Magnusson, K., Norén, F. (2014). Screening of microplastic particles in and down-stream a wastewater treatment plant. Report C55; Swedish Environmental Research Institute: Stockholm.

Mahon, A.M., O'Connell, B., Healy, M., O'Connor, I., Officer, R., Nash, R., Morrison, L. (2017). Microplastics in sewage sludge: Effects of treatment. *Environmental Science & Technology* 51, 810–818.

Mason, S.A., Garneau, D., Sutton, R., Chu, Y., Ehmann, K., Barnes, J., Fink, P., Papazissimos, D., Rogers, D.L. (2016). Microplastic pollution is widely detected in US municipal wastewater treatment plant effluent. *Environmental Pollution* 218, 1045–1054.

McCormick, A., Hoellein, T.J., Mason, S.A., Schluep, J., Kelly, J.J. (2014). Microplastic is an abundant and distinct microbial habitat in an urban river. *Environmental Science & Technology* 45 (28), 11863–11871. https://doi.org/10.1021/es503610r.

Mintenig, S.M., Int-Veen, I., Löder, M.G.J., Primpke, S., Gerdts, G. (2017). Identification of microplastic in effluents of waste water treatment plants using focal plane array-based micro-Fourier-transform infrared imaging. *Water Research* 108, 365–372.

Murphy, F., Ewins, C., Carbonnier, F., Quinn, B. (2016). Wastewater Treatment Works (WwTW) as a source of microplastics in the aquatic environment. *Environmental Science & Technology* 50(11), 5800–5808.

Nabi, I., Bacha, A.-U.-R., Li, K., Cheng, H., Wang, T., Liu, Y., … Zhang, L. (2020). Complete photocatalytic mineralization of microplastic on TiO_2 nanoparticle film. *iScience*, 101326. https://doi.org/10.1016/j.isci.2020.101326.

Napper, I.E., Bakir, A., Rowland, S.J., Thompson, R.C. (2015). Characterization, quantity and sorptive properties of microplastics extracted from cosmetics. *Marine Pollution Bulletin* 99(1–2), 178–185. https://doi.org/10.1016/j.marpolbul.2015.07.029.

Napper, I.E., Thompson, R.C. (2016). Release of synthetic microplastic plastic fibers from domestic washing machines: Effects of fabric type and washing conditions. *Marine Pollution Bulletin* 112(1–2), 39–45. https://doi.org/10.1016/j.marpolbul.2016.09.025.

Paço, A., Duarte, K., da Costa, J.P., Santos, P.S., Pereira, R., Pereira, M., Freitas, A.C., Duarte, A.C., Rocha-Santos, T.A. (2017). Biodegradation of polyethylene microplastics by the marine fungus Zalerion maritimum. *Science of the Total Environment* 586, 10–15.

Padervand, M., Lichtfouse, E., Robert, D., Wang, C. (2020). Removal of microplastics from the environment. A review. *Environmental Chemistry Letter*s. https://doi.org/10.1007/s10311-020-00983-1.

Peydayesh, M., Suta, T., Usuelli, M., Handschin, S., Canelli, G., Bagnani, M., & Mezzenga, R. (2021). Sustainable removal of microplastics and natural organic matter from water by coagulation–flocculation with protein amyloid fibrils. *Environmental Science & Technology* 55(13), 8848–8858.

Pironti, C., Ricciardi, M., Motta, O., Miele, Y., Proto, A., Montano, L. (2021). Microplastics in the environment: Intake through the food web, human exposure and toxicological effects. *Toxics*, 9, 224.

Reisser, J., Shaw, J., Wilcox, C., Hardesty, B.D., Proietti, M., Thums, M., Pattiaratchi, C. (2013). Marine plastic pollution in waters around Australia: Characteristics, concentrations, and pathways. *PLoS One* 8, https://doi.org/10.1371/journal.pone.0080466.

Rummel, C.D., Jahnke, A., Gorokhova, E., Kühnel, D., Schmitt-Jansen, M. (2017). Impacts of biofilm formation on the fate and potential effects of microplastic in the aquatic environment. *Environmental Science & Technology Letters* 4(7), 258–267.

Russell, J.R., Huang, J., Anand, P., Kucera, K., Sandoval, A.G., Dantzler, K.W., Hickman, D., Jee, J., Kimovec, F.M., Koppstein, D. (2011). Biodegradation of polyester polyurethane by endophytic fungi. *Applied and Environmental Microbiology* 77(17), 6076–6084.

Šaravanja, A., Pušić, T., Dekanić, T. (2022). Microplastics in wastewater by washing polyester fabrics. *Materials* 15(7), 2683. https://www.mdpi.com/1996-1944/15/7/2683.

Schneiderman, E.T. (2015). Discharging microbeads to our waters: An examination of wastewater treatment plants in New York. *New York State Office of the Attorney General*, 1–11. https://ag.ny.gov/sites/default/files/reports/2015_Microbeads_Report_FINAL.pdf.

Shi, X., Zhang, X., Gao, W., Zhang, Y., He, D. (2022). Removal of microplastics from water by magnetic nano-Fe_3O_4. *Science of the Total Environment* 802, 149838. https://doi.org/10.1016/j.scitotenv.2021.149838.

Sillanpaa, M., Sainio, P. (2017). Release of polyester and cotton fibers from textiles in machine washings. *Environmental Science and Pollution Research* 24 (23), 19313–19321. https://doi.org/10.1007/s11356-017-9621-1.

Sudharshan Reddy, A., Abhilash Nair, T. (2022). The fate of microplastics in wastewater treatment plants: An overview of source and remediation technologies. *Environmental Technology & Innovation* 28, 102815. https://doi.org/10.1016/j.eti.2022.102815.

Sun, J., Dai, X., Wang, Q., van Loosdrecht, M.C., Ni, B.-J. (2019). Microplastics in wastewater treatment plants: Detection, occurrence and removal. *Water Research* 152, 21–37.

Sundbæk, K.B., Koch, I., Villaro, C.G., Rasmussen, N., Holdt, S.L., Hartmann, N.B. (2018). Sorption of fluorescent polystyrene microplastic particles to edible seaweed Fucus vesiculosus. *Journal of Applied Phycology* 30(5), 2923–2927. https://doi.org/10.1007/s10811-018-1472-8.

Sutton, R., Mason, S.A., Stanek, S.K., Willis-Norton, E., Wren, I.F., Box, C. (2016). Microplastic contamination in the San Francisco Bay, California, USA. *Marine Pollution Bulletin* 109(1), 230.

Talvitie, J., Heinonen, M., Pääkkönen, J.-P., Vahtera, E., Mikola, A., Setälä, O., Vahala, R. (2015). Do wastewater treatment plants act as a potential point source of microplastics? Preliminary study in the coastal Gulf of Finland, Baltic Sea. *Water Science and Technology* 72(9), 1495.

Talvitie, J., Mikola, A., Koistinen, A., Setälä, O. (2017). Solutions to microplastic pollution: Removal of microplastics from wastewater effluent with advanced wastewater treatment technologies. *Water Research* 123, 401.

Talvitie, J., Mikola, A., Setälä, O., Heinonen, M., Koistinen, A. (2016). How well is microlitter purified from wastewater? - A detailed study on the stepwise removal of microlitter in a tertiary level wastewater treatment plant. *Water Research* 109, 164–172.

Tang, Y., Zhang, S., Su, Y., Wu, D., Zhao, Y., Xie, B. (2020). Removal of microplastics from aqueous solutions by magnetic carbon nanotubes. *Chemical Engineering Journal*, 126804. https://doi.org/10.1016/j.cej.2020.126804.

Tiwari, E., Singh, N., Monikh, F.A., Darbha, G.K. (2020). Application of Zn/Al layered double hydroxides for the removal of nano-scale plastic debris from aqueous systems. *Journal of Hazardous Materials*, 397, 122769. https://doi.org/10.1016/j.jhazmat.2020.122769.

Van Cauwenberghe, L., Devriese, L., Galgani, F., Robbens, J., Janssen, C.R. (2015). Microplastics in sediments: A review of techniques, occurrence and effects. *Marine Environmental Research* 111, 5–17. https://doi.org/10.1016/j.marenvres.2015.06.007.

Vieira, G.H., Nogueira, M.B., Gaio, E.J., Rosing, C.K., Santiago, S.L., Rego, R.O. (2016). Effect of whitening toothpastes on dentin abrasion: An in vitro study. *Oral Health and Preventive Dentistry* 14(6), 547–553. https://doi.org/10.3290/j.ohpd.a36465.

Wang, L., Kaeppler, A., Fischer, D., Simmchen, J. (2019). Photocatalytic TiO_2 micromotors for removal of microplastics and suspended matter. *ACS Applied Materials & Interfaces*. https://doi.org/10.1021/acsami.9b06128.

Watteau, F., Dignac, M.-F., Bouchard, A., Revallier, A., Houot, S. (2018). Microplastic detection in soil amended with municipal solid waste composts as revealed by transmission electronic microscopy and pyrolysis/GC/MS. *Frontiers in Sustainable Food Systems* 2, 81. https://doi.org/10.3389/fsufs.2018.00081.

Xu, Q., Huang, Q.-S., Luo, T.-Y., Wu, R.-L., Wei, W., Ni, B.-J. (2021). Coagulation removal and photocatalytic degradation of microplastics in urban waters. *Chemical Engineering Journal*, 416, 129123. https://doi.org/10.1016/j.cej.2021.129123.

Xu, X., Hou, Q., Xue, Y., Jian, Y., Wang, L. (2018). Pollution characteristics and fate of microfibers in the wastewater from textile dyeing wastewater treatment plant. *Water Science and Technology* 78(10), 2046–2054. https://doi.org/10.2166/wst.2018.476.

Yurtsever, M. (2019). Glitters as a source of primary microplastics: An approach to environmental responsibility and ethics. *Journal of Agricultural and Environmental Ethics* 32(3), 459–478. https://doi.org/10.1007/s10806-019-09785-0.

Yurtsever, M., Yurtsever, U. (2018). Use of a convolutional neural network for the classification of microbeads in urban wastewater. *Chemosphere* 216, 271–280. https://doi.org/10.1016/j.chemosphere.2018.10.084.

Zhou, G., Wang, Q., Li, J., Li, Q., Xu, H., Ye, Q., … Zhang, J. (2020). Removal of polystyrene and polyethylene microplastics using PAC and $FeCl_3$ coagulation: Performance and mechanism. *Science of the Total Environment*, 141837. https://doi.org/10.1016/j.scitotenv.2020.141837.

Ziajahromi, S., Neale, P.A., Rintoul, L., Leusch, F.D.L. (2017). Wastewater treatment plants as a pathway for microplastics: Development of a new approach to sample wastewater-based microplastics. *Water Research* 112, 93–99. https://doi.org/10.1016/j.watres.2017.01.042.

8 Characterization and Removal of Microplastics in Different Stages of Wastewater Treatment Plants

Seren Acarer Arat

8.1 SOURCES OF MICROPLASTICS IN WASTEWATER

Microplastics (MPs) are divided into two classes, primary and secondary, according to their origin (Acarer, 2023b). Sources of primary MPs in domestic wastewater are MPs in personal care products. Secondary MPs are formed as a result of the fragmentation of large plastics by physical, chemical, and biological effects. Packaging of food, beverages, personal care products and detergents, textiles, tyres and shopping bags are among the secondary sources of microplastics and secondary microplastics are formed by their breakdown. Secondary MPs in domestic wastewater consist of MPs that are significantly released into wastewater as a result of washing synthetic clothes in the washing machine. Studies have reported that 770,000–1,100,000 microfibers are released from 2 to 2.5 kg wash load of 100% polyester (PEST) t-shirts (Falco et al., 2019), and more than 700,000 fibers are released from an average of 6 kg acrylic fabric wash load (Napper & Thompson, 2016). Maintenance work carried out with paint on ships in shipyards and docks and removal and replacement of ship coatings by sanding or abrasive cleaning every few years cause high MP emissions to WWTPs or directly to the environment (Franco et al., 2020).

Domestic wastewater, industrial wastewater, rainwater, and landfill leachate containing MP are transported to WWTPs. The abundance and properties of MPs in WWTPs are closely related to the origin of the wastewater and the sewage system (Long et al., 2021). Industrial wastewaters are generally discharged after pre-treatment or directly into municipal sewerage systems. The abundance and characteristics of MPs in industrial wastewaters differ depending on the number and type of industries. For example, Franco et al. (2020) determined the MP concentration at 16 MP/L in the treatment plant of a large industrial firm that builds and repairs large ships and manufactures offshore wind structures. On the other hand, Franco et al. (2020) reported that the MP concentration in an industrial estate containing cheese-making, a slaughterhouse, a leather goods firm, a potato processing plant, marble molding,

DOI: 10.1201/9781003438793-8

and furniture manufacturing activities was 87 MP/L. In cases where combined sewer systems are used, MPs originating from tire wear together with rainwater are transported directly to WWTPs via runoff. Because leachates often contain high amounts of organic matter, inorganic salts, and toxic substances, they are treated in landfill leachate treatment plants (LLTPs) rather than in WWTPs. However, pre-treated leachate can also be discharged into the wastewater sewage system. Although leachate, which is an important source of MP, is treated with advanced treatment technologies in LLTPs, the presence of MPs in the effluent of LLTPs has been reported in some studies (Sun et al., 2021; Zhang et al., 2021). Thus, even leachate treated in LLTPs when transferred to WWTPs contributes to the abundance of MPs influent of WWTPs and affects the characteristic of MPs influent of WWTPs.

8.2 THE ABUNDANCE OF MICROPLASTICS IN WWTPS AND MICROPLASTIC REMOVAL EFFICIENCY OF WWTPS

There is no standard procedure for determining the concentration of MPs in water and wastewater. It is not possible to directly compare the results with each other, as the sampling and identification methods for the determination of MP concentration in wastewater in WWTPs differ significantly between studies. Table 8.1 shows the results of some studies reporting the MP concentration in the influent and effluent of WWTPs and the MP removal efficiency of WWTPs located in different countries. The concentration of MP in the influent of WWTPs is highly dependent on the population served by the WWTP, the consumption habits of the population, environmental awareness, industrial activities, and seasons. MPs in the influent of WWTPs are largely removed in WWTPs by going through different treatment stages such as pretreatment, primary treatment, secondary treatment, and tertiary treatment processes. The MP removal efficiency of WWTPs is highly dependent on the treatment processes applied in WWTP as well as the concentration and characteristics of the MPs in the influent. Many studies have shown that WWTPs remove more than 90% of MPs in the influent (Conley et al., 2019; Franco et al., 2021; Gies et al., 2018; Hidayaturrahman & Lee, 2019; Kwon et al., 2022; Lares et al., 2018; Xu et al., 2019). Although WWTP's MPs in the influent are largely removed from wastewater through different treatment processes, considering the capacity of WWTP, millions or billions of MP from one WWTP are discharged into the receiving environment in a day (Franco et al., 2021; Gies et al., 2018; Hidayaturrahman & Lee, 2019; Kwon et al., 2022).

MPs in conventional WWTPs are largely removed as trapped in sludge in primary and secondary settling tanks (Gies et al., 2018; Lares et al., 2018; Pittura et al., 2021). However, the use of WWTP sludge in agricultural areas causes the MPs in the sludge to spread to agricultural lands and then from soil to aquatic environments with the effect of precipitation and irrigation. Harley-Nyang et al. (2022), in WWTP, where 900 tons of anaerobically digested sludge cake and 690 tons of lime-stabilized cake are obtained per month, 1.61×10^{10} and 1.02×10^{10} MPs per month are released into the environment, respectively, by using WWTP sludges as fertilizer. They estimated that this much MP had the same equivalent volume as more than 20,000 debit cards (Harley-Nyang et al., 2022).

TABLE 8.1
The Abundance of MPs in WWTPs, MP Removal Efficiency of WWTPs, and Potential MP Release from WWTPs

WWTP Location	MP Concentration (Influent) (MP/L)	MP Concentration (Effluent) (MP/L)	Removal Efficiency	Treatment Steps	MP Release from WWTP (per Day)	Reference
USA	–	–	97.6%	Pre + P + S + T	2.91×10^8 to 5.96×10^8	Conley et al.
USA	–	–	85.2%	Pre + S + T	1.04×10^8 to 5.78×10^8	(2019)
USA	–	–	85.5%	Pre + S + T	86×10^7 to 3.08×10^8	
Spain	645	16.40	97.2%	Pre + P + S + T	$1.49–1.94 \times 10^9$	Franco et al.
Spain	1,567.4	131.3	91.6%	P + S + T	$1.07–2.64 \times 10^7$	(2021)
South Korea	4,200	33	99.2%	Pre + P + S + T	8.8×10^8	Hidayaturrahman
South Korea	31,400	297	99.1%	Pre + P + S + T	1.39×10^{11}	and Lee (2019)
South Korea	5,840	66	98.9%	Pre + P + S + T	1.37×10^9	
China	142.67	11.7	91.8%	–	–	Xu et al. (2019)
China	116	9.4	91.8%	–	–	
China	110	3.6	96.5%	–	–	
China	260.67	8.9	96.5%	–	–	
Turkey	26.5	6.9	73%	Pre + P + S	1.24×10^6	Gündoğdu et al.
Turkey	23.4	4.1	79%	Pre + P + S	3.51×10^5	(2018)
Finland	57.6	1.0	98.3%	Pre + P + S + T	1×10^7	Lares et al. (2018)
Canada	31.1	0.5	98.3%	Pre + P + S	$1 \times 10^8 - 3 \times 10^8$	Gies et al. (2018)
Korea	–	172.5	98.8%	Pre + P + S + T	8.97×10^9	Kwon et al. (2022)
	–	90	98.9%	Pre + P + S + T	2.09×10^9	Kwon et al. (2022)
	–	32	98.1%	Pre + P + S + T	1.17×10^9	Kwon et al. (2022)
Italy	2.5	0.4	84%	Pre + P + S + T	1.6×10^8	Magni et al. (2019)

Pre, pretreatment; P, primary treatment; S, secondary treatment; T, tertiary treatment.

TABLE 8.2
Dominant MP Characteristics in Influent and Effluent of WWTPs

Dominant Polymer Type (Influent)	Dominant Polymer Type (Effluent)	Dominant Shape (Influent)	Dominant Shape (Effluent)	Dominant Size (Influent + Effluent) (μm)	References
PHDA	PE	Fragment	Fiber	355–100 (influent)	Egea-Corbacho et al. (2023)
PP	PP	Granule	Granule	125–63 (influent) 355–125 (effluent)	Long et al. (2021)
PE	PU and EPM	Film	Fragment	100–500	Pittura et al. (2021)
PET	PET	Fiber	Fiber	–	Ziajahromi et al (2021)
PET	PET	Fiber	Fiber	–	Ziajahromi et al (2021)
PE	PET	Fragment	Fiber	–	Ziajahromi et al (2021)
PVC	PVC	Fiber	Fiber	355–100	Franco et al. (2020)
–	–	Fiber	Fiber	60–418	Conley et al. (2019)
–	–	Fiber	Fiber	60–418	Conley et al. (2019)
–	–	Fiber	Fiber	60–418	Conley et al. (2019)
–	–	Fragment	Fiber and fragment	20–300	Liu et al. (2019)
Rayon	Rayon	Fiber	Fiber	100–500	Xu et al. (2019)
PEST	PEST	Fiber	Fiber	1,000–5,000	Gündoğdu et al. (2018)
PEST	–	Fiber	Fiber	–	Gies et al. (2018)

EPM, ethylene-propylene-copolymer; PHDA, poly (hexadecyl acrylate), PEST, polyester.

8.3 CHARACTERISTIC OF MICROPLASTICS IN WWTPS

MPs of different polymer types, shapes, and sizes are found in WWTPs. In Table 8.2, there are results regarding the dominant polymer type, shape, and size of MPs in the influent and effluent of WWTPs reported by some studies.

8.3.1 Polymer Types

The dominant polymer types of MPs in the influent and effluent of WWTPs include polyethylene (PE), polyethylene terephthalate (PET), polystyrene (PS), polypropylene (PP), polyvinylchloride (PVC), polyamide (PA), polymethylmethacrylate (PMMA) and polyurethane (PU) is included (Franco et al., 2021; Gündoğdu et al., 2018; Pittura et al., 2021). The widespread use of these polymers in the packaging of food, beverage, detergent, personal care products, plastic bags, clothing, pipes, paints, and toys in daily life explains their presence in domestic and industrial wastewater (Acarer, 2023b). In WWTPs where domestic wastewater is treated, MPs of the PET polymer released from synthetic clothing and PE polymer type MPs found in personal care products are encountered, while MPs of various polymer types are encountered in WWTPs where industrial wastewater is treated, depending on the type of industry.

8.3.2 Shape

MPs in the influent and effluent of WWTPs exist in different shapes, such as fiber, fragment, film, bead, pellet, and foam. In the influent of WWTPs, fibers, and fragments, especially fibers, MPs are dominant (Conley et al., 2019; Gündoğdu et al., 2018; Ziajahromi et al., 2021). The origin of the fiber-shaped MPs in the influents of WWTPs is mainly the washing process of synthetic clothing. Fragmented MPs in the influent are caused by the fragmentation of large-size plastics used in daily life and plastics in industrial raw material production into smaller sizes. Due to their long and narrow shape, fiber-shaped MPs cannot be effectively retained by the treatment processes in conventional WWTPs. Fiber-shaped MPs in the effluent of WWTPs are generally more dominant than other MP shapes (Franco et al., 2021; Gündoğdu et al., 2018; Lares et al., 2018; Ziajahromi et al., 2017). Franco et al. (2021) found that 44.6% and 45.8% of MPs in the effluent of urban WWTP and industrial WWTP, where primary and secondary treatments were applied, were composed of fibers, respectively. Moreover, Ziajahromi et al. (2017) showed the dominance of fibers in wastewater treated with primary, secondary, and tertiary treatments and reverse osmosis (RO) technology. Similarly, Cai et al. (2022) reported that 94% of the MPs detected in the effluent of RO consisted of fibers.

8.3.3 Size

Generally, MPs smaller than 500 μm dominate in the influent and effluent of WWTPs (Franco et al., 2021; Pittura et al., 2021; Ren et al., 2020; Ziajahromi et al., 2017). The reason small-size MPs dominate the influents of WWTPs can be attributed to the fragmentation of MPs into smaller fragments as they are transported through the sewer system. The dominance of small-size MPs in the effluent of WWTPs is related to the better retention of large-size MPs during treatment and the fragmentation of large-size MPs into smaller pieces during treatment (Franco et al., 2020). The dominance of small-size MPs in the effluent of WWTPs highlights the need for careful monitoring of such MPs in WWTPs and the inclusion of technologies that can effectively separate small-size MPs from wastewater in conventional WWTPs.

8.4 REMOVAL OF MICROPLASTICS BY PRELIMINARY TREATMENT IN WWTPS

Table 8.3 summarizes the results of studies on MP removal efficiency of different treatment units used for pretreatment and/or primary treatment of wastewater in WWTPs. The preliminary treatment applied in WWTPs consists of grids, screens, skimming tanks, and oil and grease chamber units. Coarse screens are generally larger than 6 mm in size, while fine screens are smaller than 6 mm. By means of screens, which have a much larger opening compared to the size of MPs, MPs are retained in the screens by adhering to large particles. For instance, Blair et al. (2019) found 6% MP removal efficiency from wastewater after a coarse screen (12 mm) and grit removal process in a WWTP located in Scotland, UK. However, Ziajahromi et al. (2021) reported the removal of 69% and 79% of MPs in raw water after screening (5–6 mm) and grit removal process in two different WWTPs serving the Australian metropolitan area.

TABLE 8.3
MP Removal Efficiency of the Pretreatment and Primary Treatment Units in WWTPs

Units	Removal Efficiency	References
Grit chamber + screening	25.5%	Tadsuwan and Babel (2022)
Screening + primary settling	58.6%	Kwon et al. (2022)
Screening + primary settling	69.5%	Kwon et al. (2022)
Screening + primary settling	74.7%	Kwon et al. (2022)
Fine screening (5–6 mm) ve grit removal	69%–79%	Ziajahromi et al. (2021)
Trash rack (50 mm) + fine screening (5 mm) + grit removal	35.3%	Tadsuwan and Babel (2021)
Fine screening + grit removal + superfine screening	40.7%	Yuan et al. (2021)
Fine screening + grit removal	44.9%	Yuan et al. (2021)
Coarse screening (25 mm)	55.7%	Salmi et al. (2021)
Primary settling	47.3%	Pittura et al. (2021)
Primary settling	68.1%–88.5%	Ziajahromi et al. (2021)
Coarse (20 mm) and fine screening (6 mm) + grit removal	30%	Bilgin et al. (2020)
Coarse and fine screening + grit removal + primary settling	40.7%	Liu et al. (2019)
Coarse screening (12 mm) + grit removal	6%	Blair et al. (2019)
Grit removal + settling	56.8%–64.4%	Hidayaturrahman and Lee (2019)
Screening + grit removal + settling	41.7%	Ruan et al. (2019)
Screening + grit removal + settling	98.9%	Lares et al. (2018)
Fine screening + grit and grease removal	44.5%	Murphy et al. (2016)

Because grit chambers are designed to separate large and high-density particles from wastewater, MPs can attach to sand particles and collapse with them. The type of grit chamber used for pretreatment, such as rotary or aerated, also affects the MP removal efficiency (Lv et al., 2019), and this may cause differences between the study results. In one WWTP in China, 40.7% of MPs were removed in wastewater after fine screen, aerated grit chamber, and superfine screens, while in another WWTP in China, 44.9% were removed after fine screen and aerated grit chamber (Yuan et al., 2021). Similar results were reported in another study in Scotland, which reported removal of MPs in wastewater with 44.5% efficiency after 6 mm fine screen and grit and grease removal in a WWTP (Murphy et al., 2016). In light of the data in the literature, especially with a fine screen and grit removal pretreatment units, a significant part of the MPs in the wastewater is removed from the wastewater before going to the primary treatment.

8.5 REMOVAL OF MICROPLASTICS BY PRIMARY TREATMENT IN WWTPS

MPs in wastewater accumulate in sludge after settling in primary settling tanks and are significantly removed from wastewater (Gies et al., 2018; Pittura et al., 2021;

Ziajahromi et al., 2021). MPs, which tend to float with low density along with substances such as oil and grease, accumulate in the upper part of the tank and can be separated from the wastewater by skimming (Kwon et al., 2022). If the flow rate of the wastewater is fast in the settling tank, the sedimentation efficiency of the solid particles decreases, and the MP removal efficiency is adversely affected (Kwon et al., 2022). The amount of MP removed in the primary settling tanks, the characteristic of the MPs in the effluent of the primary settling tanks, and the characteristic of the MPs in the primary sludge depend on the concentration and characteristic of the MP in the influent. Gies et al. (2018) reported that wastewater with MP concentration of 31.1 MP/L decreased to 2.6 MP/L after primary settling, and Gies et al. (2018) reported removal of fiber and particle-shaped MPs from wastewater up to 92.8% and 88.4%, respectively, after primary settling and scum removal. Similarly, Ziajahromi et al. (2021) reported that 97.6%, 90.2%, and 80.4% of MPs were removed from the wastewater by primary treatment in three different WWTPs, and in the study, it was determined that the fibers were dominant in the primary sludge of all three WWTPs.

8.6 REMOVAL OF MICROPLASTICS BY SECONDARY TREATMENT IN WWTPS

Table 8.4 summarizes the results of some studies investigating MP removal efficiency of treatment units used for secondary treatment in WWTPs.

8.6.1 Conventional Secondary Treatment Processes

Some studies have reported that the removal efficiency of MPs from wastewater after activated sludge tank and secondary settling tank varies in the range of 14.2%–77.5% (Table 8.4). In the aeration tanks where biological treatment takes place, MP removal may not be observed in this process, as the MPs are suspended after the sludge of the secondary settling tanks is recycled (Tadsuwan & Babel, 2022). On the other hand, in secondary settling tanks following biological treatment, MPs are significantly

TABLE 8.4
MP Removal Efficiency of Secondary Treatment Units in WWTPs

Units	Removal Efficiency	Reference
Anoxic and oxic process	20%	Zhang et al. (2021)
Aeration + settling	14.2%	Tadsuwan and Babel (2021)
Aeration + settling	32%	Pittura et al. (2021)
Aeration + settling	71%	Bilgin et al. (2020)
Aeration + settling	77.5%	Ruan et al. (2019)
AAO + settling	28%	Liu et al. (2019)
MBR	50%	Zhang et al. (2021)
Anaerobic sludge blanket (UASB) + anaerobic MBR	94%	Pittura et al. (2021)
MBR	79.0%	Bayo et al. (2020)
MBR	99.9%	Talvitie et al. (2017)

trapped in sludge and separated from wastewater (Gies et al., 2018; Lares et al., 2018; Pittura et al., 2021; Tadsuwan & Babel, 2022). In a recent study by Cai et al. (2022), the MP concentration in the effluent of conventional activated sludge (CAS) decreased to 30.6 MP/L with a removal efficiency of 62.6% after primary and secondary treatment in a WWTP operated with the CAS system, but the MP concentration in the effluent was still high. In another study by Lv et al. (2019), it was found that the MP removal efficiency from wastewater in WWTP varied due to the water/sludge separation difference between the processes, and the MP removal efficiency was 15% by A/A/O process, 83.5% by membrane tank, 16.5% by oxidation ditch and 76.5% by the secondary settling tank. Since biological treatment alone does not have a great effect on MP removal from wastewater in conventional WWTPs, secondary sedimentation tanks and/or tertiary treatment units should be applied after biological treatment to remove MPs from wastewater with high removal efficiency.

8.6.2 Membrane Bioreactors

Although some studies have reported that MP removal efficiency with membrane bioreactors (MBRs) varies in the range of 50%–100% (Table 8.4), most of these studies have shown that the MP removal efficiency of MBRs is above 98% (Lares et al., 2018; Talvitie et al., 2017; Yahyanezhad et al., 2021). The differences in the properties of the membranes in MBRs, the method used for MP analysis, and the ambient conditions directly affect the amount of MP detected in the permeate of MBRs in different studies, leading to variation between the results of the studies.

MBRs remove MPs from wastewater with higher efficiency than CAS systems (Cai et al., 2022; Lares et al., 2018; Talvitie et al., 2017). The sedimentation efficiency of MPs in settling tanks following biological treatment in CAS depends on the physical properties of MPs such as size and density. On the other hand, since the separation of sludge from wastewater in MBRs depends on physical filtration, not gravity effect, MPs with larger sizes than the membrane pore sizes are effectively separated from wastewater by being retained on the membrane surface and adsorbed in the membrane pores, regardless of their density. A study to support MP removal from wastewater with higher efficiency than CAS via MBR was carried out by Di Bella et al. (2022). In their study, Di Bella et al. (2022) reported that the MP concentration in the sludge of WWTP with MBR technology was almost twice as high as the MP concentration in the sludge of two different WWTPs with the CAS process. The results showed that MPs were separated from the wastewater at a higher rate by the MBR, and there was a high amount of trapped in the sludge (Di Bella et al., 2022).

MBRs also provide MP removal with higher efficiency than the filtration processes used in tertiary treatment in WWTPs due to their superior retention efficiency. Talvitie et al. (2017) reported that the primary effluent-treating MBR can remove MPs from wastewater much more effectively due to its smaller pore size than the disc filter and sand filter used in secondary effluent-treating tertiary treatment processes. Since membranes with pore sizes corresponding to microfiltration (MF) and ultrafiltration (UF) are generally used in MBRs, a significant portion of MPs cannot pass through the membrane and accumulate in the sludge. On the other hand, some studies have shown that MPs with larger sizes than the pore sizes of the membranes used in MBRs exist in the membrane permeate (Bayo et al., 2020; Lares et al., 2018;

Zhang et al., 2021). MPs detected in the permeate of MBRs may result from possible disruption of membrane integrity, leaks between the seals in the unit, and interaction of the permeate tank with the atmosphere (Lares et al., 2018; Talvitie et al., 2017).

8.7 REMOVAL OF MICROPLASTICS BY TERTIARY TREATMENT IN WWTPS

8.7.1 Sand Filtration

MPs that are larger than the voids of the filter media cannot pass through the filter media and are separated from the wastewater. Besides sieving by size, electrostatic interaction and adsorption between filter media-MPs are other effective mechanisms in the removal of MPs by sand filtration (Hsieh et al., 2022; Talvitie et al., 2017). Studies have reported that MP removal efficiency from wastewater with sand filtration varies between 75.4% and 99.9% (Table 8.5). The abundance of MPs in the influent of the sand filter, the properties of the MPs, the properties of the filter media, and the maintenance of the filter affect the efficiency of MP removal from wastewater by sand filtration (Hsieh et al., 2022; Sembiring et al., 2021). In a recent study, it was reported that large plastic flakes from water are removed with 96.2%–99.2% efficiency by a filter media with an effective size of 0.39 mm, while smaller size tire flakes are removed with 77.8%–%99.5% efficiency by the same filter media (Sembiring et al., 2021). Some studies have shown that after sand filters, fiber and fragment-shaped MPs dominate in WWTPs (Bayo et al., 2020; Egea-Corbacho et al., 2023). Therefore, sand filtration needs to be used in combination with other treatment processes to effectively remove MPs, especially of small-size and fibers and fragment-shaped, which easily pass through the sand filter.

TABLE 8.5
Removal Efficiencies of MPs in Tertiary Treatment Units of WWTPs in Published Studies

Units	Removal Efficiency	Reference
RSF	83.1%	Egea-Corbacho et al. (2023)
RSF	99.2%–99.9%	Wolff et al. (2021)
RSF	75.4%	Bayo et al. (2020)
RSF	97.1%	Talvitie et al. (2017)
UF	75%	Zhang et al. (2021)
MF	98%	Yahyanezhad et al. (2021)
Disinfection (chlorination)	15.8%	Galafassi et al. (2022)
Disinfection (chlorination)	19.4%	
Disinfection (chlorination)	67.6%	
Disinfection (UV)	9.1%	
Disinfection (chlorination)	6.5%	Pittura et al. (2021)
Disinfection (chlorination)	−81.8%	Ruan et al. (2019)
Disinfection (chlorination)	16.7%	Liu et al. (2019)

RSF, rapid sand filter

8.7.2 Membrane Filtration

Pressure-driven membranes with decreasing pore size and increasing rising rejection efficiency are mainly categorized as microfiltration (MF), ultrafiltration (UF), nanofiltration (NF), and RO membranes, respectively. These membranes remove the MPs from wastewater via adsorption in the membrane pores and/or deposition on the membrane surface as pore diameters of the membranes are smaller than the MPs (Acarer, 2023a). Earlier studies reported that the MP removal efficiency from wastewater through pressure-driven membranes can be achieved above 94% (Pizzichetti et al., 2021; Sun et al., 2021; Yahyanezhad et al., 2021). The efficacy of the membranes' removal of MP from water and wastewater increases with a decrease in pore size (Pramanik et al., 2021). According to Pizzichetti et al. (2021), membranes composed of three distinct materials—polycarbonate (PC), cellulose acetate (CA), and polytetrafluoroethylene (PTFE)—with identical pore diameters (5 μm) but varying removal efficiencies for PA and PS MPs. Further, the efficiency of MP removal from wastewater is influenced by the characteristics of MPs such as size, charge, hydrophobicity, shape, etc., membrane specifications (porosity, pore size, charge, hydrophobicity, hardness, etc.), and interactions between membranes and MPs. Therefore, these factors must be taken into account in order to achieve high MP removal efficiencies with membranes. Membrane processes are very advantageous in that they can remove all MPs of different polymer types, shapes, and sizes from wastewater with fast operation and high efficiency. Since membrane systems are more advantageous than conventional treatment processes in removing all types of MPs from wastewater, MPs are removed from wastewater with higher efficiency with membrane systems compared to conventional treatment technologies (Cai et al., 2022).

On the other hand, during the treatment of MP-containing wastewater with membranes, two important problems come to the fore: membrane fouling and possible MP released into the treated water from the structures of polymeric membranes. Membrane fouling occurs when MPs in wastewater accumulate on the surface and pores of the membrane during filtration. With membrane fouling, the flux performance of the membrane decreases, and the transmembrane pressure increases (Enfrin et al., 2020). In particular, small-size MPs have a greater effect than large-size MPs in reducing membrane performance by increasing membrane fouling (Ma et al., 2019). In order to ensure high rejection efficiency and high flux performance, wastewater containing MP must be pre-treated before being fed to the membranes, and fouled membranes must be cleaned physically or chemically at certain periods.

Zhang et al. (2021) observed higher MP concentration in effluent than influent after the treatment of leachate with NF and RO membranes. In the study, the presence of cellulose nitrate, the polymer type of these membranes, in NF and RO effluents proved the possible loss of membrane material (Zhang et al., 2021). Moreover, a recent study by Cai et al. (2022) showed that fiber plastics with a size <200 μm can even pass through RO membranes with the smallest pore size among pressure-driven membranes, due to the smaller cross-sectional size of MPs in fiber-shaped compared to other shapes. Therefore, the effects of disruption of membrane integrity, possible openings in the system, properties of the membrane, properties of MPs, and operating conditions on MP transition from membranes need to be investigated in more detail.

8.7.3 Disinfection

Chlorination and ultraviolet (UV) irradiation are commonly used for disinfection in WWTPs. MPs react and consume the chlorine around them, and MPs degrade as a result of MP-chlorine interaction. In case of an increase in the concentration of MPs in water/wastewater, it is necessary to increase the chlorine content or extend the contact time in order to reach the targeted sterilization rate (Shen et al., 2021). As a result of contact with chlorine, bonds are damaged in the structures of MPs, MPs become rougher with deposits and cracks, and large-sized MPs break down into smaller-sized MPs. It is worth noting that these effects are different in MPs as the polymer types of MPs have different resistance to disinfectants. For instance, Kelkar et al. (2019) reported that disinfection conditions applied in drinking water treatment plants and WWTPs did not cause detectable structural changes for high-density polyethylene (HDPE) and PP, but there were detectable structural changes for PS even at low CT values (75 and 150 mg.min/L). Some studies on WWTPs have reported that the disinfection process with chlorination exhibits MP removal efficiency ranging from −81.8% to 67.6% (Table 8.5). As the degradation rate of MPs with chlorination varies depending on chlorine concentration, contact time, and temperature, the results may differ (Galafassi et al., 2022; Liu et al., 2019; Pittura et al., 2021).

UV irradiation affects the polymeric structure of MP, causing the degradation of MPs by photodegradation. Galafassi et al. (2022) stated that MPs were removed with an efficiency of 9.1% with UV irradiation in a WWTP. The disadvantages of removing MPs from wastewater by UV irradiation, which is one of the advanced oxidation processes, are the low removal efficiency and the need for long times for the degradation of MPs. Considering the current situation, convincing results are still needed to investigate the effect of disinfection with UV irradiation and chlorination on MP removal efficiency from wastewater.

8.8 CONCLUSIONS

MPs of various polymer types, sizes, and shapes, present in the influent of WWTPs, are significantly retained in different stage of wasteater treatment including preliminary, primary, secondary, and tertiary. For the investigated treatment systems, none was obseved to be removing 100% of MPs from wastewaters. This study revealed that the discharge of such untreated MPs into the receiving environment via effluent of WWTPs threatens the flora and fauna of aquatic environment. Among the existing treatment processes, membrane technologies are promising technologies with high MP removal efficiencies from wastewater. However, for the complete removal of MPs from wastewater in WWTPs, future studies should focus on eliminating the defects encountered in existing treatment processes and improving existing treatment processes.

REFERENCES

Acarer, S. (2023a). Abundance and characteristics of microplastics in drinking water treatment plants, distribution systems, water from refill kiosks, tap waters and bottled waters. *Science of the Total Environment*, *884*, 163866. https://doi.org/10.1016/j.scitotenv.2023.163866.

Acarer, S. (2023b). Microplastics in wastewater treatment plants: Sources, properties, removal efficiency, removal mechanisms, and interactions with pollutants. *Water Science & Technology*, *87*(3), 685–710. https://doi.org/10.2166/wst.2023.022.

Bayo, J., López-Castellanos, J., & Olmos, S. (2020). Membrane bioreactor and rapid sand filtration for the removal of microplastics in an urban wastewater treatment plant. *Marine Pollution Bulletin*, *156*, 111211. https://doi.org/10.1016/j.marpolbul.2020.111211.

Bilgin, M., Yurtsever, M., & Karadagli, F. (2020). Microplastic removal by aerated grit chambers versus settling tanks of a municipal wastewater treatment plant. *Journal of Water Process Engineering*, *38*, 101604. https://doi.org/10.1016/j.jwpe.2020.101604.

Blair, R. M., Waldron, S., & Gauchotte-Lindsay, C. (2019). Average daily flow of microplastics through a tertiary wastewater treatment plant over a ten-month period. *Water Research*, *163*, 114909. https://doi.org/10.1016/j.watres.2019.114909.

Cai, Y., Wu, J., Lu, J., Wang, J., & Zhang, C. (2022). Fate of microplastics in a coastal wastewater treatment plant: Microfibers could partially break through the integrated membrane system. *Frontiers in Environmental Science & Technology*, *16*(7), 96. https://doi.org/10.1007/s11783-021-1517-0.

Conley, K., Clum, A., Deepe, J., Lane, H., & Beckingham, B. (2019). Wastewater treatment plants as a source of microplastics to an urban estuary: Removal efficiencies and loading per capita over one year. *Water Research X*, *3*, 100030. https://doi.org/10.1016/j.wroa.2019.100030.

Di Bella, G., Corsino, S. F., De Marines, F., Lopresti, F., La Carrubba, V., Torregrossa, M., & Viviani, G. (2022). Occurrence of microplastics in waste sludge of wastewater treatment plants: Comparison between membrane bioreactor (MBR) and conventional activated sludge (CAS) technologies. *Membranes*, *12*, 371. https://doi.org/10.3390/ membranes12040371.

Egea-Corbacho, A., Martín-García, A. P., Franco, A. A., Quiroga, J. M., Andreasen, R. R., Jørgensen, M. K., & Christensen, M. L. (2023). Occurrence, identification and removal of microplastics in a wastewater treatment plant compared to an advanced MBR technology: Full-scale pilot plant. *Journal of Environmental Chemical Engineering*, *11*, 109644. https://doi.org/10.1016/j.jece.2023.109644.

Enfrin, M., Lee, J., Le-Clech, P., & Dum, L. F. (2020). Kinetic and mechanistic aspects of ultrafiltration membrane fouling by nano- and microplastics. *Journal of Membrane Science*, *601*, 117890. https://doi.org/10.1016/j.memsci.2020.117890.

Falco, F. De, Pace, E. Di, Cocca, M., & Avella, M. (2019). The contribution of washing processes of synthetic clothes to microplastic pollution. *Scientific Reports*, *9*, 6633. https://doi.org/10.1038/s41598-019-43023-x.

Franco, A. A., Arellano, J. M., Albendín, G., Rodríguez-Barroso, R., Quiroga, J. M., & Coello, M. D. (2021). Microplastic pollution in wastewater treatment plants in the city of Cádiz: Abundance, removal efficiency and presence in receiving water body. *Science of the Total Environment*, *776*, 145795. https://doi.org/10.1016/j.scitotenv.2021.145795.

Franco, A. A., Arellano, J. M., Albendín, G., Rodríguez-Barroso, R., Zahedi, S., Quiroga, M., & Coello, M. D. (2020). Mapping microplastics in Cadiz (Spain): Occurrence of microplastics in municipal and industrial wastewaters. *Journal of Water Process Engineering*, *38*, 101596. https://doi.org/10.1016/j.jwpe.2020.101596.

Galafassi, S., Cesare, A. Di, Nardo, L. Di, Sabatino, R., Valsesia, A., Fumagalli, F. S., Corno, G., & Volta, P. (2022). Microplastic retention in small and medium municipal wastewater treatment plants and the role of the disinfection. *Environmental Science and Pollution Research*, *29*, 10535–10546. https://doi.org/10.1007/s11356-021-16453-2.

Gies, E. A., Lenoble, J. L., Noël, M., Etemadifar, A., Bishay, F., Hall, E. R., & Ross, P. S. (2018). Retention of microplastics in a major secondary wastewater treatment plant in Vancouver, Canada. *Marine Pollution Bulletin*, *133*(June), 553–561. https://doi.org/10.1016/j.marpolbul.2018.06.006.

Gündoğdu, S., Çevik, C., Güzel, E., & Kilercioğlu, S. (2018). Microplastics in municipal wastewater treatment plants in Turkey: A comparison of the influent and secondary effluent concentrations. *Environmental Monitoring and Assessment, 190*, 626. https://doi.org/10.1007/s10661-018-7010-y Microplastics.

Harley-Nyang, D., Memon, F. A., Jones, N., & Galloway, T. (2022). Investigation and analysis of microplastics in sewage sludge and biosolids: A case study from one wastewater treatment works in the UK. *Science of the Total Environment, 823*, 153735. https://doi.org/10.1016/j.scitotenv.2022.153735.

Hidayaturrahman, H., & Lee, T.-G. (2019). A study on characteristics of microplastic in wastewater of South Korea: Identification, quantification, and fate of microplastics during treatment process. *Marine Pollution Bulletin, 146*, 696–702. https://doi.org/10.1016/j.marpolbul.2019.06.071.

Hsieh, L., He, L., Zhang, M., Lv, W., Yang, K., & Tong, M. (2022). Addition of biochar as thin preamble layer into sand filtration columns could improve the microplastics removal from water. *Water Research, 221*, 118783. https://doi.org/10.1016/j.watres.2022.118783.

Kelkar, V. P., Rolsky, C. B., Pant, A., Green, M. D., Tongay, S., & Halden, R. U. (2019). Chemical and physical changes of microplastics during sterilization by chlorination. *Water Research, 163*, 114871. https://doi.org/10.1016/j.watres.2019.114871.

Kwon, H. J., Hidayaturrahman, H., Peera, S. G., & Lee, T. G. (2022). Elimination of microplastics at different stages in wastewater. *Water, 14*, 2404. https://doi.org/10.3390/ w14152404.

Lares, M., Ncibi, M. C., Sillanpää, M., & Sillanpää, M. (2018). Occurrence, identification and removal of microplastic particles and fibers in conventional activated sludge process and advanced MBR technology. *Water Research, 133*, 236–246. https://doi.org/10.1016/j.watres.2018.01.049.

Liu, X., Yuan, W., Di, M., Li, Z., & Wang, J. (2019). Transfer and fate of microplastics during the conventional activated sludge process in one wastewater treatment plant of China. *Chemical Engineering Journal, 362*(November 2018), 176–182. https://doi.org/10.1016/j.cej.2019.01.033.

Long, Z., Wang, W., Yu, X., Lin, Z., & Chen, J. (2021). Heterogeneity and contribution of microplastics from industrial and domestic sources in a wastewater treatment plant in Xiamen, China. *Frontiers in Environmental Science, 9*, 770634. https://doi.org/10.3389/fenvs.2021.770634.

Lv, X., Dong, Q., Zuo, Z., Liu, Y., Huang, X., & Wu, W. M. (2019). Microplastics in a municipal wastewater treatment plant: Fate, dynamic distribution, removal efficiencies, and control strategies. *Journal of Cleaner Production, 225*, 579–586. https://doi.org/10.1016/j.jclepro.2019.03.321.

Ma, B., Xue, W., Hu, C., Liu, H., Qu, J., & Li, L. (2019). Characteristics of microplastic removal via coagulation and ultrafiltration during drinking water treatment. *Chemical Engineering Journal, 359*(September 2018), 159–167. https://doi.org/10.1016/j.cej.2018.11.155.

Magni, S., Binelli, A., Pittura, L., Avio, C. G., Torre, C. Della, Parenti, C. C., Gorbi, S., & Regoli, F. (2019). The fate of microplastics in an italian wastewater treatment plant. *Science of the Total Environment, 652*, 602–610. https://doi.org/10.1016/j.scitotenv.2018.10.269.

Murphy, F., Ewins, C., Carbonnier, F., & Quinn, B. (2016). Wastewater treatment works (WwTW) as a source of microplastics in the aquatic environment. *Environmental Science and Technology, 50*(11), 5800–5808. https://doi.org/10.1021/acs.est.5b05416.

Napper, I. E., & Thompson, R. C. (2016). Release of synthetic microplastic plastic fibres from domestic washing machines: Effects of fabric type and washing conditions. *Marine Pollution Bulletin, 112*(1–2), 39–45. https://doi.org/10.1016/j.marpolbul.2016.09.025.

Pittura, L., Foglia, A., Akyol, Ç., Cipolletta, G., Benedetti, M., Regoli, F., Eusebi, A. L., Sabbatini, S., Tseng, L. Y., Katsou, E., Gorbi, S., & Fatone, F. (2021). Microplastics in real wastewater treatment schemes: Comparative assessment and relevant inhibition effects on anaerobic processes. *Chemosphere, 262*, 128415. https://doi.org/10.1016/j.chemosphere.2020.128415.

Pizzichetti, A. R. P., Pablos, C., Álvarez-Fernández, C., Reynolds, K., Stanley, S., & Marugán, J. (2021). Evaluation of membranes performance for microplastic removal in a simple and low-cost filtration system. *Case Studies in Chemical and Environmental Engineering, 3*(December 2020), 100075. https://doi.org/10.1016/j.cscee.2020.100075.

Pramanik, B. K., Pramanik, S. K., & Monra, S. (2021). Understanding the fragmentation of microplastics into nano-plastics and removal of nano/microplastics from wastewater using membrane, air flotation and nano-ferrofluid processes. *Chemosphere, 282*, 131053. https://doi.org/10.1016/j.chemosphere.2021.131053.

Ren, P. J., Dou, M., Wang, C., Li, G. Q., & Jia, R. (2020). Abundance and removal characteristics of microplastics at a wastewater treatment plant in Zhengzhou. *Environmental Science and Pollution Research, 27*(29), 36295–36305. https://doi.org/10.1007/s11356-020-09611-5.

Ruan, Y., Zhang, K., Wu, C., Wu, R., & Lam, P. K. S. (2019). A preliminary screening of HBCD enantiomers transported by microplastics in wastewater treatment plants. *Science of the Total Environment, 674*, 171–178. https://doi.org/10.1016/j.scitotenv.2019.04.007.

Salmi, P., Ryymin, K., Karjalainen, A. K., Mikola, A., Uurasjärvi, E., & Talvitie, J. (2021). Particle balance and return loops for microplastics in a tertiary-level wastewater treatment plant. *Water Science & Technology, 84*(1), 89–100. https://doi.org/10.2166/wst.2021.209.

Sembiring, E., Fajar, M., & Handajani, M. (2021). Performance of rapid sand filter – single media to remove microplastics. *Water Supply, 21*(5), 2273–2284. https://doi.org/10.2166/ws.2021.060.

Shen, M., Zeng, Z., Li, L., Song, B., Zhou, C., Zeng, G., Zhang, Y., & Xiao, R. (2021). Microplastics act as an important protective umbrella for bacteria during water/wastewater disinfection. *Journal of Cleaner Production, 315*, 128188. https://doi.org/10.1016/j.jclepro.2021.128188.

Sun, J., Zhu, Z., Li, W., Yan, X., Wang, L., Zhang, L., Jin, J., Dai, X., & Ni, B. (2021). Revisiting microplastics in landfill leachate: Unnoticed tiny microplastics and their fate in treatment works. *Water Research, 190*, 116784. https://doi.org/10.1016/j.watres.2020.116784.

Tadsuwan, K., & Babel, S. (2021). Microplastic contamination in a conventional wastewater treatment plant in Thailand. *Waste Management & Research, 39*(5), 754-761. https://doi.org/10.1177/0734242X20982055.

Tadsuwan, K., & Babel, S. (2022). Microplastic abundance and removal via an ultrafiltration system coupled to a conventional municipal wastewater treatment plant in Thailand. *Journal of Environmental Chemical Engineering, 10*(2), 107142. https://doi.org/10.1016/j.jece.2022.107142.

Talvitie, J., Mikola, A., Koistinen, A., & Setälä, O. (2017). Solutions to microplastic pollution – Removal of microplastics from wastewater effluent with advanced wastewater treatment technologies. *Water Research, 123*, 401–407. https://doi.org/10.1016/j.watres.2017.07.005.

Wolff, S., Weber, F., Kerpen, J., Winklhofer, M., Engelhart, M., & Barkmann, L. (2021). Elimination of microplastics by downstream sand Filters in wastewater treatment. *Water, 13*, 33. https://doi.org/10.3390/w13010033.

Xu, X., Jian, Y., Xue, Y., Hou, Q., & Wang, L. (2019). Microplastics in the wastewater treatment plants (WWTPs): Occurrence and removal. *Chemosphere, 235*, 1089–1096. https://doi.org/10.1016/j.chemosphere.2019.06.197.

Yahyanezhad, N., Bardi, M. J., & Aminirad, H. (2021). An evaluation of microplastics fate in the wastewater treatment plants: Frequency and removal of microplastics by microfiltration membrane. *Water Pratice & Technology, 16*(3), 782–792. https://doi.org/10.2166/wpt.2021.036.

Yuan, F., Zhao, H., Sun, H., Zhao, J., & Sun, Y. (2021). Abundance, morphology, and removal efficiency of microplastics in two wastewater treatment plants in Nanjing, China. *Environmental Science and Pollution Research, 28*, 9327–9337. https://doi.org/10.1007/s11356-020-11411-w.

Zhang, Z., Su, Y., Zhu, J., Shi, J., Huang, H., & Xie, B. (2021). Distribution and removal characteristics of microplastics in different processes of the leachate treatment system. *Waste Management*, *120*, 240–247. https://doi.org/10.1016/j.wasman.2020.11.025.

Ziajahromi, S., Neale, P. A., Rintoul, L., & Leusch, F. D. L. (2017). Wastewater treatment plants as a pathway for microplastics: Development of a new approach to sample wastewater-based microplastics. *Water Research*, *112*, 93–99. https://doi.org/10.1016/j.watres.2017.01.042.

Ziajahromi, S., Neale, P. A., Silveria, I. T., Chua, A., & Leusch, F. D. L. (2021). An audit of microplastic abundance throughout three Australian wastewater treatment plants. *Chemosphere*, *263*, 128294. https://doi.org/10.1016/j.chemosphere.2020.128294.

9 Environmental Sink of Microplastics Associated with Wastewater Treatment and Allied Processes

Ansa Rebi, Taqi Raza,
Guan Wang, Muhammad Irfan, Parsa Mushtaq,
Muhammad Ali, Komal Rani Narejo,
Azfar Hussain, Muhammad Anas Khan,
and Jinxing zhou

9.1 INTRODUCTION

Our society has undergone a revolution, thanks to plastics, which has made it possible to produce a wide range of materials at low cost and for a wide range of industrial purposes. According to Thompson et al. (2009), the first three synthetic plastics were polystyrene (PS) in 1839, polyvinyl chloride (PVC) in 1835, and Bakelite in 1907. When annual worldwide plastic output reached 2 million tonnes around 1950, industrial plastic manufacturing truly started to increase. A total of 380 million tonnes of plastic are produced annually throughout the world for the manufacture of goods like storage containers, packaging, and even automobiles, and it is predicted that by the year 2050, mankind will have produced more than 30,000 million tonnes of plastic (Geyer et al., 2017).

Unfortunately, scientists have only recently realized that the qualities that make plastics helpful also make them bad for the environment. This is because plastic garbage is challenging to get rid of because it does not biodegrade in nature; instead, it just fragments into smaller bits through phytodegradation. Plastics are durable and resistant to natural deterioration due to the chemical composition that holds the molecules together. Plastics released into the environment go through processes that reduce their size, which makes managing plastic waste even more difficult. Plastics can be categorized as macro (more than 25 mm), micro (between 5 mm and 0.1 m), and nano (less than 0.1 m) based on their size.

There are several different types of microplastics, including primary and secondary microplastics. In contrast to secondary microplastics, which are created when

 DOI: 10.1201/9781003438793-9

macroplastics break down due to chemical, mechanical, photo-oxidative, and biological interactions (e.g., microfibers made from synthetic clothing), primary microplastics are intentionally formed in micro sizes for a variety of uses (e.g., microbeads in personal care products). Microplastics come in a variety of morphologies, including flakes, shafts, flakes, foams, pieces, and shafts (Su & Sakurai, 2019). Microplastics' fate and rate of environmental deterioration are determined by both their size and shape.

Microplastics have been discovered in landfills, sediments, air, and wastewater treatment facilities as a result of inappropriate waste management and plastics disposal (Klingelhöfer et al., 2020). Despite being less researched than the sources indicated above, solid waste is a substantial source of microplastic for the environment. Sludge or garbage, from wastewater treatment plants (WWTPs) and other wasted material originating from commercial or industrial activity are all considered solid waste under the Resource Conservation and Recovery Act 1976.

WWTPs may have had a significant impact on the release of microplastics into the environment. Microplastics may be removed by the WWTPs, but it depends upon unit used. However, microplastics have been demonstrated to be able to evade the WWTPs and end up in aquatic water bodies and other environment (Murphy et al., 2016). Additionally, through WWTPs effluents and sludge, microplastics may enter the environment from a variety of consumer items, such as synthetic garments, toothpaste, hand cleansers, shampoos, and shower gels due to the ineffectiveness of WWTPs. Most of the plastic trash produced by industrial activities, including non-ferrous enterprises and small workshops, is released directly into the environment without being treated (Ragaert et al., 2020). According to the United States Environmental Protection Agency, wastewater containing plastic garments can degrade and release microplastics into the environment. According to the National Academies of Sciences (Peng et al., 2021), plastic escapes into the environment due to inappropriate disposal, transportation of plastics from collection sites to a central facility, and leakage from waste management facilities. Although microplastics may not immediately kill living organisms, they might produce chronic toxicity, which is assumed to be a major problem with long-term exposure (Sussarellu et al., 2016). Because of their small size and pointed ends, microplastics have the potential to harm organisms and induce inflammation. According to research, some species that consume microplastics may experience starvation and changes to their reproductive processes (Besseling et al., 2014). Chemicals added to plastics to enhance their qualities, which are potentially hazardous to living things.

Therefore, it is important to understand all microplastic sources, as well as the associated fate and degrading mechanisms or pathways to manage microplastic pollution. Although wastewater is a significant source of microplastics in the environment, little is known about the fate, treatment, and degradation of microplastics. In this chapter, we identified the main wastewater sources that contribute to microplastics pollution, as well as their destiny, degradation, and associations with other pollutants. Future research potential is also covered.

9.2 MICROPLASTICS IN WASTEWATER

Improper use and disposal of plastics is a widespread environmental problem (Table 9.1). Microplastic does not cause any acute fatal disorder in living organisms,

TABLE 9.1
The Removal Rate of Microplastics from Wastewater and Their Retention Rate in Sludge

Country	Wastewater Treatment Stage	Microplastic Removal (%)	Microplastic Retention in Sludge	Reference
USA	Secondary	93.8	–	Michielssen et al. (2016)
USA	Tertiary	89.8	–	Michielssen et al. (2016)
USA	Tertiary	99.9	99.0%	Carr et al. (2016)
China	Secondary	90.5	–	Hohn et al. (2020)
China	Secondary	64.4	–	Liu et al. (2019)
China	Secondary	63.2	61.6%	IPCC-Jia et al. (2019)
England	Secondary	98.4	–	Murphy et al. (2016)
Canada	Secondary	98.3	93.1%	Gies et al. (2018)
Finland	Secondary	98.3	–	Lares et al. (2018)
Finland	Tertiary	99.0	80.0%	Talvitie et al. (2017)

rather causing chronic toxicity when exposure for long time. The presence of microplastic is a great public discussion because it persists in the aquatic environment and becomes part of food chain (Nor & Obbard, 2014). Above-mentioned factors are significant to reveal the potential risk about microplastics in wastewater. Microbeads, degraded plastics, and textile shedding are all ways that microplastics, or tiny pieces of plastic, get into wastewater. They are frequently not completely removed by wastewater treatment, which causes them to spread into aquatic systems. Due to its absorption by species and subsequent assimilation into the food chain, this dispersal creates ecological risks and potential health risks to people (Kumar et al., 2019). Microplastic should be banned, wastewater treatment techniques should be improved to catch microplastics, and public awareness of the negative impacts of microplastics should be raised. It is essential to address wastewater microplastics if we are to protect ecosystems and human health.

9.2.1 Sources and Types of Microplastics in Wastewater

Pharmaceutical waste can release chemicals into water bodies, while pathogenic organisms indicate contamination. Microplastics, originating from plastic breakdown, can accompany pharmaceuticals as ingredients or contaminants, affecting water quality and ecosystems. Cosmetics contain microbeads made up of walnut husk often between 0.5% and 5%, with a size up to 250 µm (Le Quéré et al., 2015). Oceans contain 35% of microplastics, which directly came from the fibers of polyester textile. Single washing of polyester fiber will produce 6 million fibers while acrylic fibers will produce 700,000 fibers (Boucher & Friot, 2017). Contact lens cleaner, glitter, small pieces of button and artificial jewelry are other sources of microplastics. Primary microplastics are present in facial and body cosmetics, pharmaceuticals, and textiles, and they can be readily released from wastewater treatment plants or transported by wind before being deposited into water resources (Gall & Thompson,

2015), while that of secondary microplastics originate from plastic waste, household plastic debris, fishing nets, and industrial wastage.

9.2.2 Fate and Behavior of Microplastics in Wastewater Treatment Plants

WWTPs are the most important collection point of microplastics in wastewater. WWTPs have diversified sources of wastewater from residential wastewater to industrial wastewater, stormwater, and commercial wastewater (Becucci et al., 2022). These several types of wastewater contain different types of microplastics including fiber, glass, fragment, sheet, and polymer. Wastewater in treatment plants also depends upon the living standards, cultural practices, ethnic values, population size, type of industry, economic conditions, season of year and location of WWTPs. Microplastics and nanoplastics entering WWTPs may discharge from the treatment plant and pollute the aquatic ecosystem (Table 9.2). Moreover, microplastics in WWTPs persist into the sludge and become part of agricultural land upon application. Furthermore, microplastic also limits treatment efficiency by slowing down the nitrification process in activated sludge and decreasing the growth of microorganisms (Li et al., 2022). Nanoplastics in WWTPs are the weathered particles of microplastics that easily escape from treatment plants and reach the soil and water reservoir. Nanoplastics are a major concern because having complex structures and large surface areas and can easily adsorb other toxic compounds upon exposure (Fang et al., 2021).

9.2.3 Effectiveness of Current Wastewater Treatment Processes in Removing Microplastics

WWTPs can degrade microplastics based on concentration and classification of microplastics. Widely applied wastewater treatments nowadays are primary, secondary, and tertiary treatments. Tertiary treatment includes ozonation, adsorption and coagulation having ability to remove microplastics up to 50%–98% (Zhu et al., 2019). Most of the solid particles are removed during grit removal or settled down by gravity. Gies et al. (2018) observed that microplastic removal efficiency was up to 92% by using primary treatment in wastewater treatment plant. In another study, Ngo et al. (2019) observed the removal efficiency of different shapes of microplastics, and granule and fragment-shaped microplastics showed removal efficiency of 91%, leading to fibers and pellets up to 79% and 83%, respectively. Another technique used for the removal of microplastics in wastewater is air flotation, in which microplastics are attached with tiny air bubbles and float onto the surface of the water, which will then be collected physically by the skimming process. Polyethylene and polypropylene are the most common examples that can be removed by air filtration technique. Secondary treatment removes microplastic up to 14%, as given in various studies (Murphy et al., 2016). Microorganisms (denitrifying bacteria) used in microplastic results in NH_3 accumulation by denitrification of treated water. Microorganisms used to remove nitrogen from water create a thin layer on the surface of microplastics. This layer, called biofilm, acts like a coating and changes how the microplastic looks and feels (Rummel et al., 2017). Moreover, activated sludge is a better option for removing microplastics than any other technique, as given by Ziajahromi et al. (2017),

TABLE 9.2
Common Plastic/Microplastic Types Found in the Wastewater and Their Toxic Effects

Type	Source	Toxicity: Ecological	Toxicity: Health Risks	References
Polyethylene terephthalate (PET)	Food and drink packaging	Chlorinated plastics can leach harmful chemicals into the soil and leak into underground water or the neighboring aquatic system	Endocrine and reproductive dysregulation, early puberty, endometriosis, and infertility	Pivnenko et al. (2015)
High-density polyethylene (HDPE)	Grocery bags, milk jugs, recycling bins, pipes	Material in pellet or bead form may mechanically cause adverse effects if ingested by waterfowl or aquatic life	Harmful effects when ingested or inhaled	Hajbane and Pattiaratchi (2017)
Polyvinyl chloride (PVC)	Door, window profiles, pipes	Creates dioxin and dioxin-like compounds, persistent environmental pollutants (POPs).	Suffer increased rates of cancer and other illnesses.	Zhang et al. (2023)
Low-density polyethylene (LDPE)	Plastic bags, straws, dispensing bottles, plastic wraps	Insoluble in water. Pellets may be harmful for birds and fish if swallowed	Irritation during eye and skin contact; Harmful effects when ingested or inhaled	Montenigro et al. (2017)
Polypropylene (PP)	Microbeads in personal care products, car parts, thermal vests	Volatile, not biodegradable and is insoluble in water	Harmful effects when ingested or inhaled	Hwang et al. (2020)
Polystyrene (PS)	Beverage cups, packing materials	Irritation of the eyes and breathing passages	Irritation during eye and skin contact	Hwang et al. (2020)

which gives removal efficiency of 66.7% and 54.4%, respectively. Generally, fragments were removed at a higher rate than fibers in secondary treatment, which suggests the difference in the shape of microplastic in wastewater. Particle size greater than 500 μm is generally removed after secondary treatment. Tertiary treatment is the last step in the removal of microplastic which can be reduced up to 0.2%–2% (Lares et al., 2018). Microplastics are removed from wastewater using a polymeric membrane during the tertiary treatment stage. Due to interactions from earlier treatments, microplastics frequently take on irregular shapes at this stage. Due to their irregular shapes, this can reduce filtration effectiveness and increase membrane wear. Up to 99% removal efficiency is achieved when using membrane bioreactors in conjunction with sand filtering. It should be noted that only fibers may easily pass through membranes lengthwise and remain in effluents. As a result, microplastics smaller than 190 m are still found in effluents after tertiary treatment, highlighting the need for cutting-edge techniques to remove fibers and nanosized particles from wastewater.

9.2.4 Efficiency of Wastewater Treatment Processes in Removing Microplastics

Advancement in the process of removing microplastics in WWTPs is the need of hour. WWTPs efficiency will be increased by advancement in the primitive technique. Currently, old techniques are used which employed primary, secondary, and tertiary techniques with some modifications. Sand filtration is an advanced technique to remove microfibers in the waste effluents having the least porosity to remove different components from wastewater. Ben-David et al. (2021) find out type, size, and shape of microplastics in domestic wastewater by using rapid sand filtration process. According to him, fibers present in the wastewater after tertiary treatment are present removed by sand filtration technique, which gives 97% removal of microplastic in wastewater. Microplastics having size of 200–400 µm size with fiber up to 80% give 64% removal after tertiary treatment. However, wastewater containing microfibers is less compared to the other microplastics in the effluents. Microplastics having a size smaller than 100 µm is removed by reverse osmosis technique. Microfibers, polystyrene, and granules up to 99% are easily removed by reverse osmosis after tertiary treatment technique. Advanced techniques include membrane technology with anti-fouling ability and having low operational cost, less capital cost, easy to use and maximum sustainability to adverse operating conditions (Liu et al., 2022). Flow rate is all time challenge in membrane technology, which can be reinforced by making composite membrane. Eukaryotic marine plants, such as seagrasses, are the advanced method that acts as a barrier against microplastics and helps in sludge treatment. They can also be helpful by sequestering carbon, as a source of food and habitat for different organisms.

9.2.5 Ecological and Human Health Impacts of Microplastics Release from Wastewater Treatment Plants

Plastic waste accumulation is an emerging cause of concern associated with biodiversity and human health. Most of the studies related to plastic waste concluded that coastal areas are more exposed to microplastic than open seas which suggests that anthropogenic activation is an important factor behind all this scenario. WWTPs are another important source of plastic waste which shows point source pollution (Li et al., 2020). Microplastics are capable of absorbing heavy metals and pharmaceutical waste and may persist in the environment due to strong hydro stability of microplastic in environment. Microplastics have the potential to become part of human food chain by ingestion of sea and terrestrial food. Sea salt is another major source of microplastics. The concentration of microplastics in Spanish and Chinese table salt is 50–280 and 7–680 microplastics/kg. PET is a potential human carcinogen is found in plastic film, drink bottles, insolation molding, microwave packaging, etc. PS and PVC additives were used in packaging, disposable cutlery, and other plastic items, which may lead to cancer and reproductive disorders, especially in humans, rodents, and invertebrates. In human beings, it may accumulate in the blood. Microplastics in air taken by inhalation process have adverse effects on human health, such as particle toxicity, pathogen, and parasitic attack. Polystyrene nanoplastic particles were found in the blood and other organs of rats usually translocated from the gastrointestinal tract. Phthalate esters used in different materials help increase flexibility and longevity, are harmful for humans upon exposure, and cause disorders in birth and unusual sexual growth. Potentially harmful human health concerns are associated with microplastic toxicity, which not only absorbs the environment but also can cause toxicity in the environment.

9.3 MICROPLASTICS IN SLUDGE

9.3.1 Fate and Behavior of Microplastic in Sludge

The presence, transformation, and fate of microplastics in sewage sludge have received less attention than in other water bodies. Therefore, the presence, fate, and behavior of microplastics are important to focus. Microplastics can enter to any component of the environment after entering the ecosystem from any source. For instance, current treatment and disposal techniques may lead to microplastics in food waste, sludge, and trash from landfills, eventually ending up in the same location (such as a water body or agricultural soil). Large amounts of microplastics are retained in sewage sludge and wastewater because of the low effectiveness with which small plastic particles are removed in the majority of WWTPs (Nizzetto et al., 2016).

The majority of the microplastics found in sludge and wastewater are discharged by personal care items, synthetic textile fibers, microbeads, and residential washing machine effluent. The removal of microplastics is determined by the effectiveness of various treatment units in WWTPs, and this efficiency varies depending on the physicochemical and biological processes used. According to research, early treatment methods including skimming and screening can remove 35%–58% of microplastics from wastewater (Murphy et al., 2016), and the sewage sludge effectively preserved up to 99% of the microplastics present in raw sewage. The concentrations of sludge microplastics vary geographically, probably because of various population densities and waste management techniques. For instance, Finland reported 301.4 items/kg of sludge, whereas China reported 271,700 items/kg. Contrary to countries with sparse populations, those with higher population densities consume more personal care products, generate more laundry wastewater, and send more wastewater-to-wastewater treatment facilities. Most investigations have revealed that microplastics retained in sludge are often greater in size than those in wastewater. This is explained by the possibility of bigger microplastics sinking to the tank's bottom while smaller microplastics are still moving through the WWTPs (Murphy et al., 2016).

9.3.2 Potential Risks and Impacts of Sludge Disposal and Application on Soil and Environment

Nations now view sludge as a resource rather than a waste due to the ongoing production of enormous amounts of sewage sludge and the useful components found in it (Raza et al., 2023). The primary recycling method of sludge is land application as a soil amendment due to its high organic and nutritional value (Nizzetto et al., 2016). Even though applying sludge improves soil fertility, it also opens a conduit for pathogens and, more crucially, the accumulation of some trace pollutants (Figure 9.1), such as toxic organics and microplastics, into agricultural soils. Due to the varying compositions, characteristics, and additions of microplastics, there may be a variety of ecological threats to both agricultural soils and human life. It is obvious that WWTPs contribute to an additional channel for microplastic contamination in the environment, given that over 90% of the microplastics in wastewater are collected in sludge (Li et al., 2020). Therefore, it is noted that the main source of microplastics entering the environment is the use of sludge in agriculture. Sludge amendment in the soil is a significant mechanism for microplastics to enter the soil from sludge. According to studies, irrigation systems are allowing enormous amounts of microparticles from wastewater to reach agricultural fields. More than half of the entire amount of produced sludge is used as agricultural fertilizer in Canada, the United States, and Portugal. According to Yang and Zhang (2015), more than 80% of the treated sludge in China is dumped in landfills.

The effects of microplastics on agricultural soils can be attributed to several factors, including the fact that ingesting microplastics may slow down earthworm growth, reduce nematode survival and body length, and slow down collembolan growth. The digestive system clogs created by microplastics in the fauna's gut, which reduce food intake and nutrient absorption, or even injuries to the skin and digestive tract brought on by some sharp microplastics (Shaw et al., 2023). Recent studies have documented how microplastics affect terrestrial plants in addition to soil animals.

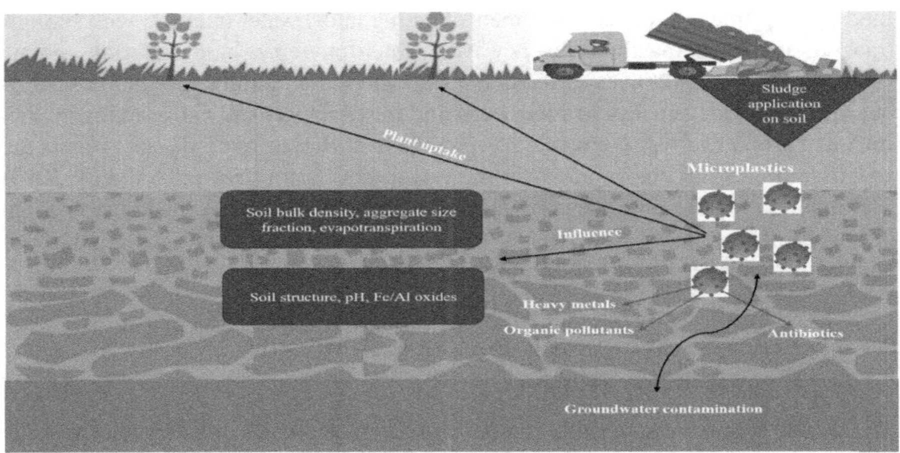

FIGURE 9.1 Sludge application on soil.

Microplastics can alter soil parameters such as soil aggregate size fraction, soil bulk density, and evapotranspiration, in addition to their direct effects on crops and soil animals. By expanding the routes for water transport, microplastics may hasten soil water evaporation and alter the microbial population of the soil, particularly the root-colonizing microorganisms.

According to studies, several metals, including Cd, Mn, Pb, As, Cu, Zn, and Cr, can adsorb on the surface of microplastics (Abbasi et al., 2020). High-density polyethylene was also found to be linked to As, Cd, Cr, and Pb. Since groundwater is frequently exposed to metals that are either geogenic or anthropogenic in origin, their interaction with microplastics is unavoidable. Thus, in addition to microplastics themselves, metals and metalloids that have been absorbed on them also end up in food that is consumed by humans and animals. Metals are known to have harmful effects on human health (Raza et al., 2021).

9.4 MICROPLASTICS IN RECEIVING WATER BODIES

9.4.1 Transport and Fate of Microplastics in Receiving Water Bodies

The presence of microplastics in receiving water bodies has sparked alarm due to the possible environmental and ecological consequences. Microplastics have been detected in the water columns and sediments of lakes and rivers across the world. The amount and mass of plastic waste eroding through a river may be larger than that of living organisms such as zooplankton and fish larvae. The bioavailability of microplastics to marine animals rises as their number increases. Color, density, shape, size, charge, aggregation, and abundance of these microscopic plastic particles all have an impact on their potential bioavailability to marine creatures (Wright et al., 2013). Many freshwater invertebrates (e.g., "shredders" like gammarid amphipod) play important roles in the breakdown of particulate organic matter and may interact with microplastics as well. Birds, fish, bivalves, crustaceans, and other invertebrates ingest microplastics and can transfer them through aquatic and terrestrial food webs. Large microplastic fragments floating in the water interfere with sunlight transmission, reducing the efficiency with which phytoplankton undertake photosynthesis. Phytoplankton is an essential biological community in aquatic settings because it provides energy to food webs and performs critical ecosystem activities such as carbon cycling. However, the predicament due to microplastic depreciates as most sufferers go unexplored over the vast oceans (Pabortsava & Lampitt, 2020). Ingression of plastics into the ecosystem is mainly due to erroneous human actions or unrestrained wastes from water or sewage treatment plants and textile industries, and numerous other creatures' activities, for example, chewing.

9.4.2 Impacts of Microplastic Pollution on Aquatic Ecosystems and Wildlife

The health of aquatic species is significantly impacted when floating microplastics are mistakenly consumed as food. Microplastics enter aquatic systems by waste discharge from treatment facilities, overflow of sewers during heavy rains, and biosolid runoff from agricultural areas (Issac & Kandasubramanian, 2021). Therefore, the uptake of microplastic by aquatic biota is a severe concern (Raza et al., 2022a, b). Although microplastic accumulation was found in the fishes' intestines, it has more serious ramifications when consumed. As a result, it is safe to rule out the possibility of microplastic transmission by eating. The presence of microplastic in the digestive tract of dried fish, however, can pose a major risk to human health because the fish are consumed without first being gutted. One method of human uptake of microplastics is even the intake of whole or undeveined dried white shrimp. Other species including the edible oyster, big clam, and green mussel have also raised similar safety concerns. Beyond just aquatic and terrestrial habitats, microplastics have an influence on animals. Further, plastics are easily used to trap and entangle wildlife, making it difficult for them to move about in search of food or making them more vulnerable to neighboring predators.

9.5 TECHNOLOGIES AND STRATEGIES FOR REMOVING MICROPLASTICS FROM WASTEWATER

9.5.1 Physical Separation Techniques

Physical separation techniques play a vital role in the removal of microplastics from wastewater. These techniques include processes such as sedimentation, filtration, and centrifugation, which rely on the size, density, and surface properties of microplastics for their separation. Advancements in filtration media, such as membrane filters and granular media filters, have shown promising results in removing microplastics from wastewater. Understanding the efficiency, limitations, and optimization of physical separation techniques is crucial for their successful implementation in microplastic pollution control.

9.5.2 Advanced Oxidation Processes

Advanced oxidation processes (AOPs) offer potential solutions for the degradation of microplastics in wastewater (Kim et al., 2022). Techniques such as ozonation, photocatalysis, and Fenton processes utilize chemical reactions to break down microplastics into smaller, less harmful fragments. AOPs can target specific types of microplastics, and organic pollutants associated with them. However, optimizing process conditions, energy requirements, and the potential formation of by-products are important considerations when employing AOPs for microplastic removal.

9.5.3 Biological Treatment Methods

Biological treatment methods offer sustainable approaches for microplastic removal from wastewater. Biofiltration, activated sludge processes, and constructed wetlands can enhance microbial degradation and bioaccumulation of microplastics. The microbial community present in wastewater treatment systems can play a crucial role in breaking down microplastics and associated organic compounds. Understanding the microbial ecology, optimizing operational parameters, and integrating biological treatment methods into wastewater treatment plants are essential for effective microplastic removal (Priyadharshini et al., 2021).

9.5.4 Adsorption Techniques

Adsorption techniques have shown promise in removing microplastics from wastewater. Activated carbon, biochar, and other adsorbents can attract and capture microplastics through surface interactions. These techniques can be applied in batch or continuous-flow systems and can be combined with other treatment processes for enhanced efficiency. Optimization of adsorbent properties, regeneration methods, and scale-up considerations are important aspects to consider when utilizing adsorption techniques for microplastic removal (Kumar et al., 2023).

9.6 CHALLENGES AND FUTURE DIRECTIONS

Addressing the challenges associated with microplastic pollution control requires ongoing research and innovation. Developing cost-effective and scalable technologies, understanding the fate and behavior of microplastics in wastewater treatment processes, and assessing the long-term ecological impacts are key areas for future investigation. Collaborative efforts are importantly required among researchers and industries, stakeholders, including policymakers, manufacturers, and consumers. Governments can implement policies and regulations that promote the use of alternative materials, incentivize sustainable practices, and support research and development in the field. Policies and regulations on microplastic pollution should be subject to periodic review and adaptation to keep pace with scientific advancements and emerging challenges. Governments need to establish mechanisms for regular policy evaluation, incorporating new scientific findings and addressing gaps or limitations in existing regulations.

REFERENCES

Abbasi, S., Ayoob, T., Malik, A., & Memon, S. I. (2020). Perceptions of students regarding E-learning during Covid-19 at a private medical college. *Pakistan Journal of Medical Sciences, 36*(COVID19-S4), S57.

Becucci, M., Mancini, M., Campo, R., & Paris, E. (2022). Microplastics in the florence wastewater treatment plant studied by a continuous sampling method and Raman spectroscopy: A preliminary investigation. *Science of the Total Environment, 808*, 152025.

Ben-David, I., Franzoni, F., Moussawi, R., & Sedunov, J. (2021). The granular nature of large institutional investors. *Management Science, 67*(11), 6629–6659.

Besseling, E., Wang, B., Lurling, M., & Koelmans, A. A. (2014). Nanoplastic affects growth of S. obliquus and reproduction of D. magna. *Environmental Science & Technology, 48*(20), 12336–12343.

Boucher, J., & Friot, D. (2017). *Primary Microplastics in the Oceans: A Global Evaluation of Sources* (Vol. 10): Iucn Gland, Switzerland.

Carr, S. A., Liu, J., & Tesoro, A. G. (2016). Transport and fate of microplastic particles in wastewater treatment plants. *Water Research, 91*, 174–182.

Fang, C., Sobhani, Z., Zhang, X., McCourt, L., Routley, B., Gibson, C. T., & Naidu, R. (2021). Identification and visualisation of microplastics/nanoplastics by Raman imaging (III): Algorithm to cross-check multi-images. *Water Research, 194*, 116913.

Gall, S. C., & Thompson, R. C. (2015). The impact of debris on marine life. *Marine Pollution Bulletin, 92*(1–2), 170–179.

Geyer, R., Jambeck, J. R., & Law, K. L. (2017). Production, use, and fate of all plastics ever made. *Science Advances, 3*(7), e1700782.

Gies, E. A., LeNoble, J. L., Noël, M., Etemadifar, A., Bishay, F., Hall, E. R., & Ross, P. S. (2018). Retention of microplastics in a major secondary wastewater treatment plant in Vancouver, Canada. *Marine Pollution Bulletin, 133*, 553–561.

Hajbane, S., & Pattiaratchi, C. B. (2017). Plastic pollution patterns in offshore, nearshore and estuarine waters: A case study from Perth, Western Australia. *Frontiers in Marine Science, 4*, 63.

Hohn, S., Acevedo-Trejos, E., Abrams, J. F., de Moura, J. F., Spranz, R., & Merico, A. (2020). The long-term legacy of plastic mass production. *Science of the Total Environment, 746*, 141115.

Hwang, J., Choi, D., Han, S., Jung, S. Y., Choi, J., & Hong, J. (2020). Potential toxicity of polystyrene microplastic particles. *Scientific Reports, 10*(1), 1–12.

IPCC-Jia, G., Shevliakova, E., Artaxo, P., De Noblet-Ducoudré, N., Houghton, R., House, J., ... & Verchot, L. (2019). Land–climate interactions. *Climate Change and Land: an IPCC special report on climate change, desertification, land degradation, sustainable land management, food security, and greenhouse gas fluxes in terrestrial ecosystems.* https://www.ipcc.ch/site/assets/uploads/2019/11/05_Chapter-2.pdf

Issac, M. N., & Kandasubramanian, B. (2021). Effect of microplastics in water and aquatic systems. *Environmental Science and Pollution Research, 28*, 19544–19562.

Kim, S., Sin, A., Nam, H., Park, Y., Lee, H., & Han, C. (2022). Advanced oxidation processes for microplastics degradation: A recent trend. *Chemical Engineering Journal Advances, 9*, 100213.

Klingelhöfer, D., Braun, M., Quarcoo, D., Brüggmann, D., & Groneberg, D. A. (2020). Research landscape of a global environmental challenge: Microplastics. *Water Research, 170*, 115358.

Kumar, S., Prasad, S., Yadav, K. K., Shrivastava, M., Gupta, N., Nagar, S., & Yadav, S. (2019). Hazardous heavy metals contamination of vegetables and food chain: Role of sustainable remediation approaches: A review. *Environmental Research, 179*, 108792.

Kumar, V., Singh, E., Singh, S., Pandey, A., & Bhargava, P. C. (2023). Micro-and nano-plastics (MNPs) as emerging pollutant in ground water: Environmental impact, potential risks, limitations and way forward towards sustainable management. *Chemical Engineering Journal, 459*, 141568.

Lares, M., Ncibi, M. C., Sillanpää, M., & Sillanpää, M. (2018). Occurrence, identification and removal of microplastic particles and fibers in conventional activated sludge process and advanced MBR technology. *Water Research, 133*, 236–246.

Le Quéré, C., Moriarty, R., Andrew, R. M., Canadell, J. G., Sitch, S., Korsbakken, J. I., & Boden, T. A. (2015). Global carbon budget 2015. *Earth System Science Data, 7*(2), 349–396.

Li, L., Luo, Y., Li, R., Zhou, Q., Peijnenburg, W. J., Yin, N., & Zhang, Y. (2020). Effective uptake of submicrometre plastics by crop plants via a crack-entry mode. *Nature Sustainability, 3*(11), 929–937.

Li, S., Ding, F., Flury, M., Wang, Z., Xu, L., Li, S., & Wang, J. (2022). Macro-and microplastic accumulation in soil after 32 years of plastic film mulching. *Environmental Pollution, 300*, 118945.

Liu, X., He, P., Chen, W., & Gao, J. (2019). Improving multi-task deep neural networks via knowledge distillation for natural language understanding. *arXiv preprint arXiv:1904.09482.*

Liu, Y., Wang, B., Pileggi, V., & Chang, S. (2022). Methods to recover and characterize microplastics in wastewater treatment plants. *Case Studies in Chemical and Environmental Engineering, 5*, 100183.

Michielssen, M. R., Michielssen, E. R., Ni, J., & Duhaime, M. B. (2016). Fate of microplastics and other small anthropogenic litter (SAL) in wastewater treatment plants depends on unit processes employed. *Environmental Science: Water Research & Technology, 2*(6), 1064–1073.

Montenigro, P. H., Alosco, M. L., Martin, B. M., Daneshvar, D. H., Mez, J., Chaisson, C. E., & Cantu, R. C. (2017). Cumulative head impact exposure predicts later-life depression, apathy, executive dysfunction, and cognitive impairment in former high school and college football players. *Journal of Neurotrauma, 34*(2), 328–340.

Murphy, F., Ewins, C., Carbonnier, F., & Quinn, B. (2016). Wastewater treatment works (WwTW) as a source of microplastics in the aquatic environment. *Environmental Science & Technology, 50*(11), 5800–5808.

Ngo, P. L., Pramanik, B. K., Shah, K., & Roychand, R. (2019). Pathway, classification and removal efficiency of microplastics in wastewater treatment plants. *Environmental Pollution, 255*, 113326.

Nizzetto, L., Futter, M., & Langaas, S. (2016). Are agricultural soils dumps for microplastics of urban origin? *Environmental Science and Technology ASAP* (20). doi: 10.1021/acs.est.6b04140.

Nor, N. H. M., & Obbard, J. P. (2014). Microplastics in Singapore's coastal mangrove ecosystems. *Marine Pollution Bulletin, 79*(1–2), 278–283.

Pabortsava, K., & Lampitt, R. S. (2020). High concentrations of plastic hidden beneath the surface of the Atlantic Ocean. *Nature Communications, 11*(1), 4073.

Peng, Y., Wu, P., Schartup, A. T., & Zhang, Y. (2021). Plastic waste release caused by COVID-19 and its fate in the global ocean. *Proceedings of the National Academy of Sciences, 118*(47), e2111530118.

Pivnenko, K., Jakobsen, L., Eriksen, M. K., Damgaard, A., & Astrup, T. F. (2015). *Challenges in plastics recycling.* Paper presented at the Proceedings of the Sardinia.

Priyadharshini, S. D., Babu, P. S., Manikandan, S., Subbaiya, R., Govarthanan, M., & Karmegam, N. (2021). Phycoremediation of wastewater for pollutant removal: A green approach to environmental protection and long-term remediation. *Environmental Pollution, 290*, 117989.

Ragaert, K., Huysveld, S., Vyncke, G., Hubo, S., Veelaert, L., Dewulf, J., & Du Bois, E. (2020). Design from recycling: A complex mixed plastic waste case study. *Resources, Conservation and Recycling, 155*, 104646.

Raza, T., Ali, I., Khan, N., Eash, N. S., Qadir, M. F., & Jatav, H. S. (2023). Detoxification of sewage sludge by natural attenuation and application as a fertilizer. In *Environmental Pollution Impact on Plants* (pp. 245–270). Apple Academic Press. doi: 10.1201/9781003304210-10.

Raza, T., Qureshi, K. N., Imran, S., Eash, N. S., & Bortone, I. (2021). Associated health risks from heavy metal-laden effluent into point drainage channels in Faisalabad, Pakistan. *Pakistan Journal of Agricultural Research, 34*(3), 487–494.

Raza, T., Shehzad, M., Abbas, M., Eash, N. S., Jatav, H. S., Sillanpaa, M., & Flynn, T. (2022a). Impact assessment of COVID-19 global pandemic on water, environment, and humans. *Environmental Advances*, 11, 100328.

Raza, T., Shehzad, M., Qadir, M. F., Kareem, H. A., Eash, N. S., Sillanpaa, M., & Hakeem, K. (2022b). Indirect effects of COVID-19 on water quality. *Water-Energy Nexus*, 5, 29–38.

Rummel, C. D., Jahnke, A., Gorokhova, E., Kühnel, D., & Schmitt-Jansen, M. (2017). Impacts of biofilm formation on the fate and potential effects of microplastic in the aquatic environment. *Environmental Science & Technology Letters, 4*(7), 258–267.

Shaw, V., Mondal, A., Mondal, A., Koley, R., & Mondal, N. K. (2023). Effective utilization of waste plastics towards sustainable control of mosquito. *Journal of Cleaner Production, 386*, 135826.

Su, J., Vargas, D. V., & Sakurai, K. (2019). One pixel attack for fooling deep neural networks. *IEEE Transactions on Evolutionary Computation, 23*(5), 828–841.

Sussarellu, R., Suquet, M., Thomas, Y., Lambert, C., Fabioux, C., Pernet, M. E. J., & Epelboin, Y. (2016). Oyster reproduction is affected by exposure to polystyrene microplastics. *Proceedings of the National Academy of Sciences, 113*(9), 2430–2435.

Talvitie, J., Mikola, A., Koistinen, A., & Setälä, O. (2017). Solutions to microplastic pollution–Removal of microplastics from wastewater effluent with advanced wastewater treatment technologies. *Water Research, 123*, 401–407.

Thompson, R. C., Swan, S. H., Moore, C. J., & Vom Saal, F. S. (2009). *Our Plastic Age* (Vol. 364, pp. 1973–1976): The Royal Society Publishing, London.

Wright, S. L., Thompson, R. C., & Galloway, T. S. (2013). The physical impacts of microplastics on marine organisms: A review. *Environmental Pollution, 178*, 483–492.

Yang, J., & Zhang, Y. (2015). Protein structure and function prediction using I-TASSER. *Current Protocols in Bioinformatics, 52*(1), 5.8.1–5.8.15.

Zhang, Q.-Q., Yu, Y., Liu, J.-Z., Fu, W.-J., Quan, J.-Y., Chen, Y., & Jin, R.-C. (2023). Evaluation the role of soluble microbial products for denitrification sludge characteristic under starvation stress. *Science of the Total Environment, 882*, 163319.

Zhu, C., Li, D., Sun, Y., Zheng, X., Peng, X., Zheng, K., & Mai, B. (2019). Plastic debris in marine birds from an island located in the South China Sea. *Marine Pollution Bulletin, 149*, 110566.

Ziajahromi, S., Neale, P. A., Rintoul, L., & Leusch, F. D. (2017). Wastewater treatment plants as a pathway for microplastics: Development of a new approach to sample wastewater-based microplastics. *Water Research, 112*, 93–99.

10 Mass Balances and Life Cycle of Microplastics Across Wastewater Treatment Plant

A Process Review

Dharmesh H. Sur and Divya O. Tirva

10.1 INTRODUCTION

Global plastic production reached 370 million metric tons in 2021. That also indicates the proportionate quantum of pollution. Needless to say, though, plastics have dominantly secured their place through a variety of domestic and industrial products. Pollution created through these plastic products imposes serious concerns on the environment as it is non-degradable in nature and durable too. Microplastics (MPs) are very tiny microparticles with a size of less than 5 mm. These (MPs) are detrimental to water bodies and ecosystems. Living beings of water bodies or affected ecosystems consume these MPs of smaller size. Due to their smaller size, they cannot be easily identified or separated in comparison to the larger size of plastics. These MPs contaminate the environment and, hence, become toxic. The oceans and other water bodies constitute an ecosystem with zooplankton, phytoplankton, and algae. MPs interact with these biotic components, affecting carbon and oxygen cycles. MPs in the soil may also have an impact on the structure and characteristics of the soil, plant performance, and microbial activity. The health effects of MPs are reviewed by Lee et al. (2023). Hydrophobic contaminants deposit on the surface of MPs since they are hydrophobic.

Reddy and Nair (2022) reviewed the origin of MPs in wastewater. This includes their makeup, remediation from wastewater treatment units, and harmful effects. Primary, secondary, and tertiary stages of wastewater treatment make unique contributions to MPs towards its treatment. The residual content of MPs in the effluent is significant and demands a need to treat it in all its stages through primary, secondary, and tertiary wastewater treatments. According to published research, more than 90% of MPs may be separated from wastewater by treatment plants. Physicochemical properties and characteristics of the source-generating MPs need to be studied since the risk posed by MPs in a biotic (as well as abiotic) environment is critical. The terrain ecosystems are no exception when it comes to pollution caused by microplastics.

 DOI: 10.1201/9781003438793-10

It raises a significant concern about sensitivity to the environment. There is a need to identify the exposure levels of such microplastics to the environment and make plans for it. The plans to identify the sources of such microplastics along with its sinks, can save terrestrial ecosystems (Harley-Nyang et al., 2022).

Removal or separation and degradation technologies make up the categories of remediation technologies for microplastic removal from wastewater. Additionally, the sources, fate, and movement of microplastics via the inlets and outlets of wastewater can be examined. Last but not least, suggestions have been made for the future in order to solve MP pollution in wastewater treatment plants and, ultimately, lessen the strain on the aquatic and soil ecosystems. This chapter emphasises the clear possibilities of various processes through primary, secondary, and tertiary wastewater treatments.

10.2 MASS BALANCE

Rasmussen et al. (2021) have taken into account all sizes entering the facility, and the overall plastic mass removal efficiency was 99.6%. When solely taking into account the modest MP of 500 m, the removal efficiency rose to 98.8%. The sludge retained 65.8% of the total, followed by the 2 mm bar screens with 11.6% of the total retained, and the effluent with 1.2%. The main process tanks and digester caused the degradation, in part. The error caused due to measurement and sampling in the same might have resulted in up to 21.4% of disparity, which is higher than expected. This mass balance had a disparity of 21.4% that was thought to be the result of measurement and sampling errors or degradation in the parts like main process tanks as well as digester. The 73% of the total load was made up of the material that was kept on the screen. In their work, much larger polymers were retained by the bar screen material. These larger polymers comprise other than existing in effluent, influent, and sludge. This study demonstrates that it is essential to examine all the components of plastics including microplastics. The work also emphasises on taking into account all sizes of plastics at all treatment phases, like primary, secondary, and tertiary treatments. This approach shall help in establishing the extensive mass balance idea or related concepts.

10.2.1 Primary Treatment

Gravity sedimentation is a common wastewater treatment method that removes settleable solids from screened and gritted wastewater. Typically, little more than half of the suspended solids are removed in primary treatment as far as municipal wastewater treatment is concerned. Additionally, biological oxygen demand (BOD) in the form of sediment particles is removed. It is recommended to provide a settlement period for microplastics. Microplastics are fairly settled, and they further need to be concentrated for their removal (Figure 10.1a).

Primary treatment is more effective in removing microplastics prior to its secondary treatment. This step favours the economy in operation too. The primary treatment's waste product (primary sludge) contains the removed elements (nutrients, pathogenic organisms, trace elements, and potentially harmful organic compounds including microplastics) and is now treated in secondary treatment (10%–20%

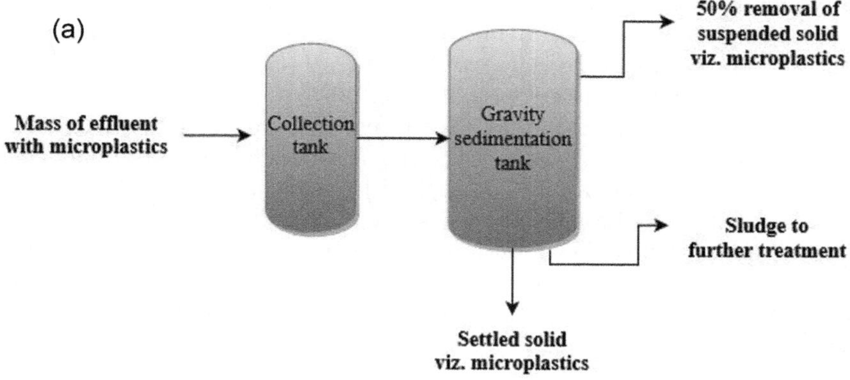

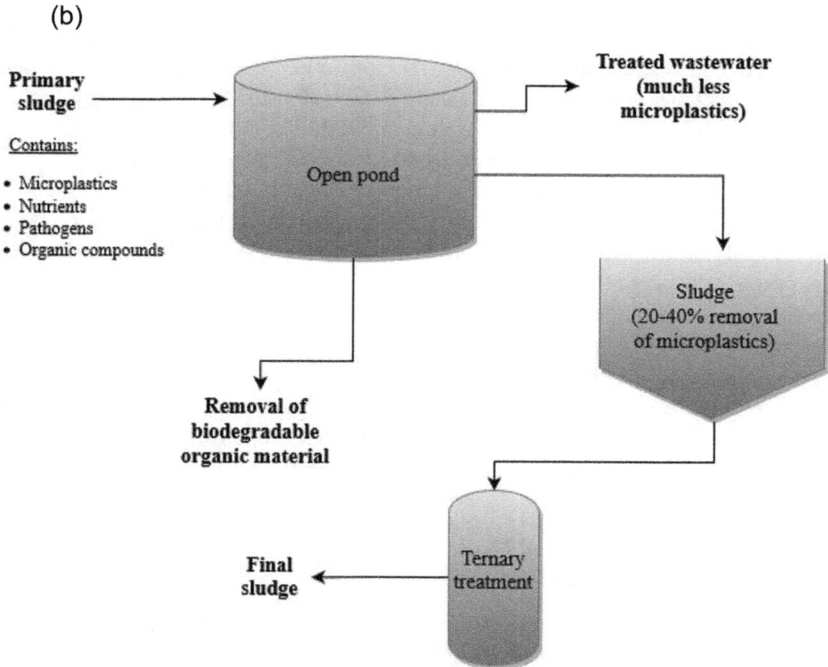

FIGURE 10.1 (a) Mass balance schematic diagram for microplastics during primary treatment and (b) secondary and tertiary treatments.

removal of microplastics) and left fractions to the tertiary treatment, as seen in Figure 10.1a. The microorganisms undergo metabolic activity and flocculate. They form settleable particles. This process follows the removal of secondary sludge (excess biomass after secondary treatment) in sedimentation tanks. Here also it is recommended to provide a separate mass balance calculation towards the settled mass of microplastics.

Lofty et al. (2022) have worked for microplastics, mentioning the role of the primary settler tank. The Primary settler tank (PST) removes microplastics with the size ranging from 1,000 to 5000 μm from the inflow to the PST and separated by density into settled sludge and surface scum, which provided 17% and 4% of microplastics by mass as well as its quantum, respectively. Sludge and scum are mixed to create sewage sludge, which contains 0.01 g MP or 24.7 MPs per gram of dried sludge, or roughly one per cent of the total weight of sludge being discharged. Authors favoured the opinion that soils serve as a big reservoir that can sequester microplastics because MPs are transferred from wastewater treatment plants to agricultural land, highlighting the severity of the MP pollution problem resulting from the direct recycling of sewage sludge into organic fertiliser.

10.2.2 Secondary Treatment

As discussed in the previous section, the sewage sludge contains the amount of microplastics (as mentioned: 0.01 g MP or 24.7 MPs per gram of dried sludge), and the removal percentage of the same treatment is 40% around by the end of secondary treatment (Figure 10.1b). Sludge treatment is recommended in aerobic mode and in ponds as well. The rate of aeration is the vital parameter here. The sludge drying period shall justify the final amount of microplastics retained in the secondary sludge treatment. A possibility for the treatment of secondary sludge with soil is suggested, which may capture some amount of microplastics at that stage. This step makes it easy to calculate the amount of microplastics. It is proposed here to treat this amount of microplastics with the microorganisms with species like *Alcaligenes faecalis* and *Bacillus cereus*, as they showed the highest degradation activity (Tareen et al., 2022).

10.2.3 Tertiary Treatments

Characteristics of input wastewater like domestic wastewater, industrial effluents, and stormwater runoff, would decide the extent of microplastics retained until tertiary treatment. Major mass of microplastics, more than up to 50%, is treated before this stage. The sludge remained from secondary treatments (aerobic process, microbial, and soil remediation) enters the reaction tank meant for tertiary treatment in its full volume, at a time, with the whole mass including microplastics to the maximum of 20%–30% untreated volume. This volume of mass needs to be settled. The processes like filtration, sedimentation as well as chemical processes like coagulation and flocculation can be taken into action, in their effective combination. It is also interesting to note that effect of ultraviolet rays on plastics can bring degradation from 8 to 640 days (Cheng et al., 2021). This UV effect can be added to existing or designed treatment process systems. The sludge can be stabilised. Stabilised sludge with organic solids is called biosolids. When the biosolids are recycled, they become a substantial source of microplastic contamination, as seen on the plains, as far as the United Kingdom alone is concerned. The amount of recycled biosolids every year from the wastewater industry is 3.5 million tonnes. In order to address this problem, the current investigation by Harley-Nyang et al. (2022) identified the presence of microplastics throughout the entire system of sludge treatment at a model wastewater treatment facility in the United

Kingdom. Both biosolids (sewage sludge that has been treated) and sewage sludge (a liquid byproduct) were sources of microplastics in all wastewater treatment procedures. All samples collected from various stages of the treatment process contained microplastics (37.7–286.5 microplastics/g of sludge (dry weight)). This study has referred to the production of digested cake of sludge as well as lime-stabilised cake every month at a given wastewater treatment plant. Lime-stabilised sludge receives 1.02×10^{10} amount of microplastics as against digested sludge (anaerobic) that facilitates the release of 1.61×10^{10} amount of microplastics when biosolids are used as fertilisers. The findings demonstrated how harmful microplastics may be. The presence of microplastics in sludge and biosolids is not currently governed by any laws or guidelines.

There exists the concept of retention ponds also for microplastics treatments. Liu et al. (2019) worked on runoff from urban and roadway surfaces, those responsible for discharging pollutants during storm occurrences, which comprise microplastics to almost considerable amount. Liu et al. (2019) quantified such ponds and investigated for the composition, its form, size as well as mass of microplastics and concluded on organic matter content, hydraulic load and catchment features. The ideas on proportions of microplastics in organic matters, hydraulic load and catchment area (or volume) decide the parameters while designing wastewater treatment for total mass of microplastics to be treated. The surface-to-volume ratio of the mass of microplastics and its degradation capacity fixes the untreated mass of microplastics. Microplastic quantity, the composition of the polymer, size distribution, the catchment's land use, and the amount of organic matter in the sediment did not show any link with one another. However, there was a link between the hydraulic loading that is received by the ponds as well as the amount of microplastics present in their sediments. This study demonstrated that some of the microplastics can be trapped by sediments in stormwater retention ponds, preventing their downstream movement. When analysing the destiny of microplastics from urban and highway environments, these systems must be taken into account.

Kim et al. (2022) worked on season variations to investigate its effects on tertiary treatment. The abundance of microplastics in the incoming streams to the wastewater treatment plants (influents) showed notable seasonal changes. The significant seasonal variations were displayed by the quantity of microplastics entering the systems; however, no particular trend was identified in the effluents during seasons, indicating that the treatment plant has a consistent performance (approx. 99.8%). Spatial assessments of microplastics distribution in the treatment plant revealed that while coagulation-sedimentation and quick sand filtering removed the majority of microplastics, some microplastics continued to exist in the treated effluent to yield a potential yearly load of 2.9×10^9 particles and 0.54 kg into the rivers. These findings emphasise the significance of long-term monitoring and MP mass measurement, which would allow for more precise calculation of microplastic load.

The seasonal distribution of microplastics as well as the quantity entering into the aquatic environment, were investigated in a study by Kim et al. (2022) in a tertiary treatment plant. Despite the fact that the microplastics abundance in the influent fluctuated seasonally, the wastewater treatment plant used tertiary procedures, such as fast sand filtration, to remove more than 99% of the microplastics in all seasons. The average microplastics size going through quick sand filtration was reduced considerably, resulting in the presence of microplastics larger than 50m in the effluent received finally. This highlighted that more accurate quantification of microplastics discharge would require the development of reliable, consistent sampling procedures that catch small-sized microplastics. Furthermore, rainfall greatly contributed to the mass loading of microplastics. Thus, using the collected data on seasonal samples and the mass data of microplastics, one can improve the accuracy of microplastics loading predictions.

10.3 DISTRIBUTION AND OCCURRENCE OF MICROPLASTICS

Van Do et al. (2022) suggested that an evaluation of the occurrence of microplastics inside wastewater treatment plants be conducted along with its distribution. The findings may help to improve the managerial issues and subsequent action plans of treatment plants for the control of microplastics contamination (or its prevention) before discharge into the environment. The majority of microplastics found in this study had a profile of their shape, colour, size, and chemical composition. With the plentiful MPs kinds of fibre and fragment in both the influent and the effluent, the sizes of microplastics ranged widely from 1.6 to 5,000 mm.

10.4 CHALLENGES

Microplastics pollution resulting from wastewater treatment plants is reviewed for challenges, detection, and sustainable removal techniques by Sadia et al. (2022). Microplastics are largely removed ineffectively by microbial breakdown. With no secondary pollution load, the membrane reactor is quite effective, but surface fouling and plastic particles that collect in the pores cause a considerable decline in effectiveness. They are not sustainable, inefficient, expensive, and could cause secondary pollution. It is still difficult to get microplastics out of contaminated aquatic habitats. The combination of various techniques for microplastics removal includes solar energy, UV (ultraviolet) applications, 3D solar evaporators, photocatalysis, etc, and to some extent – the aerobic processes. These approaches are fairly straightforward. They all promises to remove the microplastics from the wastewater effectively and cheaply. A significant percentage of MPs originates from our daily activities, therefore, future efforts may focus on creating household-scale microplastic processing equipment that could aid in preventing microplastic contamination at the source. It is, therefore, recommended not to process much of microplastic mass in aerobic or anaerobic wastewater treatments alone. Primary, secondary, and tertiary treatments suggested in this chapter describe the removal of microplastics by 10%–20%, 20%–40%, and more than 40%–50%, respectively. The combination of conventional processes with nonconventional ways like UV, solar, etc. can be implemented.

10.5 CHARACTERISE MICROPLASTICS IN WASTEWATER TREATMENT PLANTS

Wastewater treatment plants can have the potential to recover and characterise microplastics (Liu et al., 2022). While the aforementioned approaches and techniques have been shown to be efficient for the identification of microplastics in the samples collected, additional findings are still required to develop dependable, more approachable, and user-friendly approaches for the analysis of microplastics in wastewater treatment facilities. First off, there is still an opportunity for improvement in the removal of materials (both organic and inorganic) from the wastewater and sludge samples. The removal of inorganic and organic materials from sludge samples, as well as secondary treatment processes, needs attention. The solution, in part, to this problem can be hydrolysis and enzymatic treatment. For example, in wastewater treatment plants, where the salts of iron and aluminium are used for the purpose of removing phosphorous, the plant contains metal oxides after treatment. These metal-organic precipitates cannot be completely removed after their second or third run of wastewater treatment. As mentioned in the first paragraph of "challenges," the majority of the microplastic waste comes from domestic activities. Therefore, the break in household generations remains in focus for the future. Mitigating such waste remains an immediate priority to curb the mass of microplastics. This is discussed in the next point.

10.6 STRATEGIES FOR MITIGATION OF MICROPLASTIC MASS

Mitigating microplastic pollution requires a comprehensive approach that addresses various sources, pathways, and impacts. Here are some strategies, along with relevant references.

10.6.1 Reducing Single-Use Plastics

Implementing bans or restrictions on single-use plastics can significantly diminish the input of debris of plastics released into the environment. According to the Plastic Waste Management Amendment Rules, 2021, the Indian government designated the following products as single-use plastics with poor usefulness and high littering potential. As a result, they are forbidden: plastic or PVC banners less than 100 µm, cutlery which includes wrapping films, straws, non-reusable straws, forks, plates, cigarette packets, cards made for events and invitations, plastic flags, thermocol for decoration, cups, stirrers, glasses, sticks used in balloons, ice creams, candy, ear-buds, etc. Additionally, as of September 30, 2021, the notification forbids the production, importation, stocking, distribution, sale, and use of plastic carry bags with a thickness of less than 75 µm, and as of December 31, 2022, it forbids a thickness of less than 120 µm.

Pandey (2023) reports that the average annual consumption of plastic per person worldwide is 20.9 kg, compared to 5.3 kg in India. According to Swiss-based research consultancy – Earth Action (EA), Plastic Overshoot Day, which is observed on July 28, 2023 (the day on which the amount of plastics exceeds the global waste management capacity), highlights a crucial aspect of plastic consumption worldwide: short-life plastics, which include plastic packaging and single-use plastics. These categories make up about 37% of all plastic that is commercialised each year. They also present a greater danger of environmental leakage. This year, an additional 68.64 Megatons of plastic trash will find its way into the environment, according to the 2023 Plastic Overshoot Day Report by EA. Among the 12 nations accountable for 52% of the world's improperly handled plastic garbage is India. One of these is microplastics.

The amount of land-based plastic garbage that enters the ocean was approximated by Jambeck et al. (2015) by tying together global data on solid waste, population density, and economic status. According to one of their data sets, 192 coastal countries produced 275 million metric tons (MT) of plastic debris in 2010, of which 4.8–12.7 million MT found their way into the ocean. Studies on the size of the population and the effectiveness of waste management systems are necessary to determine the detrimental role that plastics, especially microplastics, play in the formation of debris.

10.6.2 Improving Waste Management

Improving the infrastructure for garbage disposal, recycling, and collection aids in keeping plastic out of rivers. The marine environment attracts a lot of microplastics from many sources. Andrady (2011) discusses the potential causes and ecological effects of these microplastics, which have become contaminants in the marine environment. Andrady (2011) came to the conclusion that the fracturing and surface embrittlement of plastics in beach environments due to weathering appears to be a major mechanism for microplastic formation. It is highly likely that the poisons caused by microplastics will go widely and through trophic levels via food chains and the food web.

10.6.3 Innovations in Packaging

Developing and promoting sustainable packaging alternatives can help reduce the generation of plastic waste. There are several sustainable packaging alternatives to traditional plastic that aim to reduce environmental impact. Here are some options:

1. Biodegradable Plastics
    - PLA (Polylactic Acid): Compostable in industrial settings, PLA is derived from renewable resources such as corn starch, as well as sugarcane.
    - PHA (Polyhydroxyalkanoates): Produced by bacteria, PHA is fully biodegradable and can be used for various packaging applications.
2. Compostable Packaging

    Compostable Bags: Made from materials like cornstarch, compostable bags break down into natural elements when disposed of in a composting environment.

    Mushroom Packaging: The root structure of mushrooms – mycelium – can be utilised to make compostable and biodegradable packaging materials.
3. Paper-Based Packaging

    Cardboard: A widely used and recyclable material, cardboard is a good alternative for various packaging applications.

    Paper Foam: Made from recycled paper fibres, paper foam is a lightweight and protective packaging material.

4. Plant-Based Plastics

   PHA and PLA (mentioned above): Plant-based plastics are derived from renewable resources and can be used in various applications.

   Bioplastics: These are made from agricultural by-products or non-food crops like sugarcane.
5. Glass

   Glass Bottles and Jars: Glass is a sustainable material for food and drink packaging since it is infinitely recyclable and reusable.
6. Metal

   Aluminium: Lightweight and easily recyclable, aluminium is often used for beverage cans and food packaging.
7. Textile Packaging

   Fabric Bags: Reusable fabric bags can replace single-use plastic bags for packaging purposes.
8. Edible Packaging

   Edible Films: Some companies are exploring edible packaging made from materials like seaweed or starch that can be consumed along with the food.
9. Recycled Plastics

   Post-Consumer Recycled (PCR) Plastics: Using recycled plastics helps reduce the demand for new plastic production.
10. Hybrid Packaging

    Combination Materials: Some companies are experimenting with combining different materials to create packaging that is both durable and sustainable. When considering sustainable packaging alternatives, it is essential to assess the entire life cycle of the material, including its sourcing, production, use, and disposal. Additionally, local recycling capabilities and infrastructure play a crucial role in the effectiveness of sustainable packaging solutions.

In their discussion of the ultimate destination of all plastics generated, Geyer et al. (2017) revealed that by 2050, almost 12,000 Mt of plastic garbage will be in landfills or the environment, if present production and waste management patterns continue.

10.6.4 Filtration and Screening Technologies

Installing effective filtration systems in wastewater treatment plants can capture microplastics before they reach aquatic environments. One of the simplest methods is using mesh filters with fine pore sizes to physically trap microplastics. In a study published in 2021, Akarsu et al. (2021) looked at the removal of MPs from wastewaters using membrane filtration and electrocoagulation-electro flotation (EC/EF) techniques. To achieve optimal treatment efficiency, the effects of several electrode combinations (Fe-Al and Al-Fe), current density (10–20 A/m^2), pH (4.0–10.0), and working periods (0–120 minutes) on the removal of two distinct polymer particles in water were studied. In terms of microplastic removal, the suggested procedures offered the highest removal efficiency (100%) in contrast to traditional secondary and tertiary treatment techniques (2%–81.6%). Two distinct urban and sub-urban locations were used to study the atmospheric fallout of microplastics (between 2 and 355 particles/m^2/day) (Dris et al., 2016). A stainless-steel funnel was used to continuously gather microplastics. After filtering, the samples were examined. So, this process can also be replicated in filed after its customised parameters are realised to help decrease microplastics pollution. Biological filtration can also be facilitated to ingest microplastics. As experimented by Scherer et al. (2017), all species (*Lumbriculus variegatus*, *Physella acuta*, *Gammarus pulex*, *Chironomus riparius*) ingested microplastics in a concentration-dependent manner.

10.7 THE ROLE OF REGULATIONS, PUBLIC AWARENESS, AND TECHNOLOGICAL SOLUTIONS

The role of regulations, public awareness, and technological solutions is crucial in addressing this issue. Here's an overview of each component.

10.7.1 Regulations

Ban or Restriction on Microplastics: Governments can enact regulations to ban or restrict the use of certain types of microplastics in consumer products, such as personal care products and cleaning agents. This can reduce the source of microplastics entering the environment. Waste Management Policies: Improved waste management policies can help prevent the release of microplastics from larger plastic items. Regulations regarding proper disposal and recycling practices are essential to minimise the fragmentation of plastics into microplastics. Industrial Standards: Regulations can be implemented to encourage industries to adopt environmentally friendly production processes and materials, reducing the overall generation of microplastics during manufacturing.

10.7.2 Public Awareness

Education Campaigns: Public awareness campaigns can inform individuals about the sources, impacts, and proper disposal methods of plastics. These campaigns can motivate consumers to make environmentally conscious choices and reduce their use of plastic products. Behavioural Change: A greater understanding may result in altered consumer behaviour, such as choosing products with less plastic packaging, supporting companies with sustainable practices, and participating in community clean-up initiatives. Advocacy: Informed and engaged citizens can advocate for stronger regulations and policies at local, national, and international levels, putting pressure on governments and industries to take action against microplastics.

10.7.3 Technological Solutions

Filtration and Waste Water Treatment: Advanced wastewater treatment technologies can be employed to capture and remove microplastics from wastewater before it is released into natural water bodies. This includes the development of filtration systems and other treatment methods designed to specifically target microplastics.

Innovative Materials: The creation of substitute materials that are readily recyclable or biodegradable can aid in lowering the environmental persistence of microplastics. Research into environmentally friendly packaging materials and substitutes for goods made of microplastics is part of this.

Sensor Technologies: Microplastics can be tracked and found in a variety of situations by using advanced sensor technologies. Finding the sources of pollution and evaluating the efficacy of mitigation strategies can be made easier with the help of this sensor-collected data.

Effective mitigation of microplastics requires a combination of regulatory frameworks, public engagement, and technological innovations. Collaboration between governments, industries, researchers, and the public is essential to develop and implement comprehensive strategies to address this environmental issue.

10.8 CONCLUSIONS

The removal of microplastics is possible to a considerable extent in a wastewater treatment plant. Solid's removal in primary treatment decides the sludge quantities being carried away through secondary and tertiary treatment steps. Mass balance across each tank in primary treatment is easier since it involves only sedimentation of clarifying processes mainly. It is going to be a crucial and indicative step as far as further process requirements for total microplastic removal are concerned. Secondary treatment reduces microplastics further by 40%–50% of its total entering load. The majority of solids are removed during the primary settling processes. The mass entering secondary treatment can have two choices to be operated at – aerobic and anaerobic. Aerobic mode of operation helps remove microplastics with ease. However, this treatment step is recommended to combine with other nonconventional methods like UV, solar, etc., as mentioned in the chapter. The activated sludge does not contribute to treating primary sludge except by flocculating some of the suspended solids, including microplastics (10%–20% removal only). However, the tertiary sludge facilitates the excess removal of microplastics by quantity (20%–40%) after it was significantly accomplished during the primary treatment (50 % removal).

ACKNOWLEDGEMENTS

The author is grateful to Prof. (Dr.) R.B. Jadeja (Dean-Research, Marwadi University, Rajkot, Gujarat, India) for his motivation towards the research.

REFERENCES

Akarsu, C., Kumbur, H., & Kıdeyş, A. E. (2021). Removal of microplastics from wastewater through electrocoagulation-electroflotation and membrane filtration processes. *Water Science and Technology*, 84(7), 1648–1662. https://doi.org/10.2166/wst.2021.356.

Andrady, A. L. (2011). Microplastics in the marine environment. *Marine Pollution Bulletin*, 62(8), 1596–1605. https://doi.org/10.1016/j.marpolbul.2011.05.030.

Cheng, F., Zhang, T., Liu, Y., Zhang, Y., & Qu, J. (2021). Non-negligible effects of UV irradiation on transformation and environmental risks of microplastics in the water environment. *Journal of Xenobiotics*, 12(1), 1–12. https://doi.org/10.3390/jox12010001.

Dris, R., Gaspéri, J., Saad, M., Mirande, C., & Tassin, B. (2016). Synthetic fibers in atmospheric fallout: A source of microplastics in the environment? *Marine Pollution Bulletin*, 104(1–2), 290–293. https://doi.org/10.1016/j.marpolbul.2016.01.006.

Geyer, R., Jambeck, J., & Law, K. L. (2017). Production, use, and fate of all plastics ever made. *Science Advances*, 3(7). https://doi.org/10.1126/sciadv.1700782.

Harley-Nyang, D., Memon, F. A., Jones, N., & Galloway, T. (2022). Investigation and analysis of microplastics in sewage sludge and biosolids: A case study from one wastewater treatment works in the UK. *Science of the Total Environment*, 823, 153735. https://www.statista.com/statistics/282732/global-production-of-plastics-since-1950/, June 2023.

Jambeck, J., Geyer, R., Wilcox, C., Siegler, T. R., Perryman, M. E., Andrady, A. L., Narayan, R., & Law, K. L. (2015). Plastic waste inputs from land into the ocean. *Science*, 347(6223), 768–771. https://doi.org/10.1126/science.1260352.

Kim, M. J., Na, S. H., Batool, R., Byun, I. S., & Kim, E. J. (2022). Seasonal variation and spatial distribution of microplastics in tertiary wastewater treatment plant in South Korea. *Journal of Hazardous Materials*, 438, 129474.

Lee, Y., Cho, J., Sohn, J., & Kim, C. (2023). Health effects of microplastic exposures: Current issues and perspectives in South Korea. *Yonsei Medical Journal*, 64(5), 301.

Liu, F., Vianello, A., & Vollertsen, J. (2019). Retention of microplastics in sediments of urban and highway stormwater retention ponds. *Environmental Pollution*, 255, 113335.

Liu, Y., Wang, B., Pileggi, V., & Chang, S. (2022). Methods to recover and characterize microplastics in wastewater treatment plants. *Case Studies in Chemical and Environmental Engineering*, 5, 100183.

Lofty, J., Muhawenimana, V., Wilson, C. A. M. E., & Ouro, P. (2022). Microplastics removal from a primary settler tank in a wastewater treatment plant and estimations of contamination onto European agricultural land via sewage sludge recycling. *Environmental Pollution*, 304, 119198.

Pandey, K., India among the 12 countries responsible for 52% of the world's mismanaged plastic waste: Report, Down to Earth, 31 July 2023.

Rasmussen, L. A., Iordachescu, L., Tumlin, S., & Vollertsen, J. (2021). A complete mass balance for plastics in a wastewater treatment plant-macroplastics contributes more than microplastics. *Water Research*, 201, 117307.

Reddy, A. S., & Nair, A. T. (2022). The fate of microplastics in wastewater treatment plants: An overview of source and remediation technologies. *Environmental Technology & Innovation*, 28, 102815.

Sadia, M., Mahmood, A., Ibrahim, M., Irshad, M. K., Quddusi, A. H. A., Bokhari, A., & Show, P. L. (2022). Microplastics pollution from wastewater treatment plants: A critical review on challenges, detection, sustainable removal techniques and circular economy. *Environmental Technology & Innovation*, 28, 102946.

Scherer, C., Brennholt, N., Reifferscheid, G., & Wagner, M. (2017). Feeding type and development drive the ingestion of microplastics by freshwater invertebrates. *Scientific Reports*, 7(1). https://doi.org/10.1038/s41598-017-17191-7.

Tareen, A., Saeed, S., Iqbal, A., Batool, R., & Jamil, N. (2022). Biodeterioration of microplastics: A promising step towards plastics waste management. *Polymers*, 14(11), 2275. https://doi.org/10.3390/polym14112275.

Van Do, M., Le, T. X. T., Vu, N. D., & Dang, T. T. (2022). Distribution and occurrence of microplastics in wastewater treatment plants. *Environmental Technology & Innovation*, 26, 102286.

11 Wastewater Treatment Plants as a Key Source of Secondary Microplastic in the Urban Environment

A Case Study from Uttarakhand, India

Manish Chaudhary and Surindra Suthar

11.1 INTRODUCTION

Plastics have become ubiquitous in our everyday lives due to their many useful qualities, including durability, affordability, lightness, flexibility, non-reactivity, low thermal and electrical conductivity, and corrosion resistance. Globally, 8.3 million tonnes of plastic were produced, according to (*Plastic Europe*, 2019). Subsequently, the content of waste plastic has been increasing enormously in municipal solid waste and wastewater streams in the recent past. Only 9% of plastic waste is recycled, 12% is burned, and 79% ends up in landfills, dumps, and oceans, resulting in environmental contamination. Mismanagement and plastic durability lead to prolonged environmental persistence, lasting hundreds of years, with most plastic never fully decomposing. However, large plastic waste materials gradually fragment into smaller pieces through processes like photo-, thermos-, or bio-degradation. These minute plastic particles, measuring less than 5 mm in diameter, are termed as microplastics (MPs). MP currently raises global concerns due to potential adverse impacts on wildlife and ecosystem quality. Their characteristics, including extensive presence in the human environment (Yu et al., 2020), toxic attributes (such as butyl phthalate, polybrominated diphenyl ethers, and synthetic pigments) (Ren et al., 2020), absorptive properties (primarily due to their substantial surface area and hydrophobic nature) (Gatidou et al., 2019), bioaccumulation tendencies, and prolonged environmental persistence, classify MPs as an emerging pollutant.

Primary MPs refer to large plastic items designed for specific purposes such as drug delivery systems, cosmetic abrasives, engineering, and industrial applications

DOI: 10.1201/9781003438793-11

(Auta et al., 2017). Secondary MPs can also be unintentionally generated through various human activities. For example, residential washing machines release a significant amount of microfibers into the environment, averaging 1,472–2,121 microfibers per garment. Additionally, tyre wear, vehicle transportation, road markings, and brakes are significant sources of MPs, contributing approximately 0.81 kg per person. Plastic pellets, fishing gear, construction paints, and other sources release significant amounts of MPs into surface water bodies each year (Vivekanand et al., 2021). Urban runoff is another significant source of MPs in aquatic environments, and meteorological conditions such as heavy precipitation, strong winds, and flooding can worsen MP pollution (Fischer et al., 2016). The use of plastic mulch in agriculture also contributes to MP deposition in soils (Wang et al., 2020). MPs can enter the atmosphere through various sources, including emissions from vehicles, industries, and aerosols. Wind also plays a role in dispersing MPs in the atmosphere from terrestrial surfaces (Abbasi et al., 2019). Study conducted in Greater Paris reported an average deposition rate of 236 particles/m^2/day from the atmosphere (Dris et al., 2016).

WWTPs are intended to collect and treat various forms of household or industrial wastewater and are recognized as one of the major point sources of dumping MPs into the environment (Long et al., 2019; Talvitie et al., 2017) (see Figure 11.1). About 90% of MP content is removed in WWTPs; consequently, substantial amounts of MPs are discharged into rivers and oceans through the disposal of treated effluents from such WWTPs (Edo et al., 2020). MPs recovered from treated water are predominantly found in sludge. The abundance of MPs in SS collected from WWTPs has been documented extensively across various regions globally (see Table 11.1). Larger MPs tend to settle with sludge during primary and secondary settling processes, while smaller

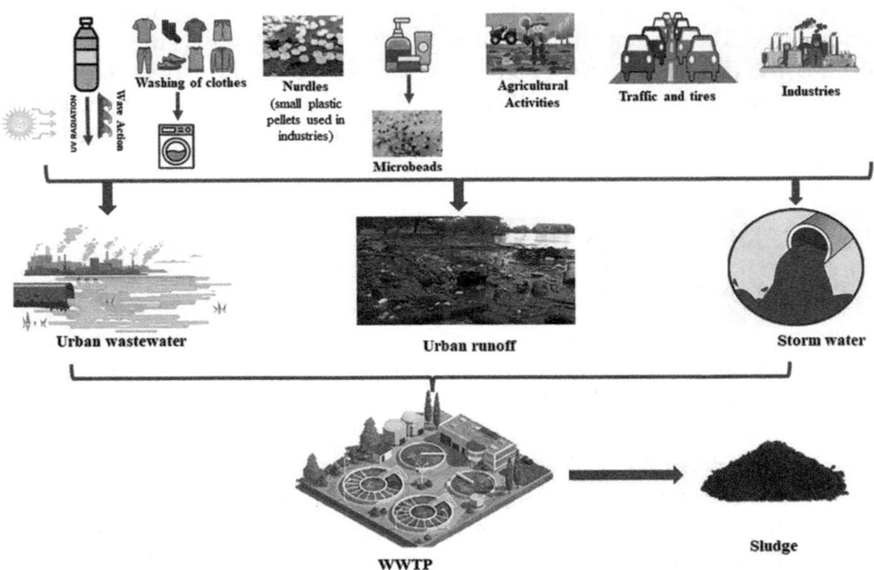

FIGURE 11.1 Sources of MP in the sewage sludge

TABLE 11.1
Comparing MP Concentrations in Sewage Sludge in Different Parts of the World

Sample	Country	MP Size Range	MP Concentration (Particles/kg)	Reference
Sewage sludge	Swedish	300 μm–5 mm	$16.7 \pm 1.96 \times 10^3$	Magnusson and Norén (2014)
Sewage sludge	Germany	10 μm–5 mm	1,000–24,000	Lassen et al. (2015)
Sewage sludge	Netherland	0.7 μm–5mm	370–950	Brandsma et al. (2013)
Sewage sludge	China	37 μm–5 mm	1,565–56,386	Li et al. (2018)
Sewage sludge	USA	<5 mm	Five but all the particles were below 1%	Carr et al. (2016)
Sewage sludge	Ireland	250 μm–4 mm	4,196–15,385	Mahon et al. (2017)
Sewage sludge	USA	-	About 1–4 particles/gram	Zubris and Richards (2005)
Sewage sludge	Uttarakhand, India	20 μm–5 mm	29,200–36,600	This study

particles may remain in the water or become absorbed in the sludge during the treatment process in WWTPs (Rolsky et al., 2020). SS is managed through landfill disposal, incineration, or soil application, with the latter being the most common method in Europe, North America, and South Asia (Hurley et al., 2020; Rolsky et al., 2020).

This research assessed the quantity and characterization of MPs in SS from two wastewater treatment plants (WWTPs) situated in Uttarakhand, India. SS samples were collected from two wastewater treatment plants (WWTPs) utilizing distinct technologies. These samples were then examined to determine the presence of microplastics (MP) based on the number of particles/kg of sludge, their shapes, their size distribution, and their colour. The outcomes of this study hold the potential to serve as a foundational basis for establishing a comprehensive policy framework to mitigate MP pollution across the Ganga River basin.

11.2 METHODOLOGY

11.2.1 Study Area

The SS samples were obtained from two separate WWTPs in Uttarakhand, India, which largely receive wastewater from households of Dehradun and Rishikesh city. The technology of selected WWTPs assures that wastewater treatment fulfils the most recent environmental regulations. Table 11.2 provides comprehensive insights of factors like WWTP inlet load, treatment processes, sludge generation, and the assigned nomenclature for sampling locations. Selected cities are densely populated and well-known for religious tourism and rapid economic growth. Dehradun is the capital with the highest population density in the Indian state of Uttarakhand. Rishikesh, situated alongside the Ganga River, attracts millions of visitors annually for religious purification baths as per Hindu traditions and for adventurous activities.

TABLE 11.2
Description of WWTPs

Sample Name	Province	WWTP Location	City	STP Capacity	Sludge Generation (m^3/day)	Treatment Technology
SS-A	Uttarakhand	Kargi Road, Morowala	Dehradun	68 MLD	7.59 (approx.)	SBR
SS-B		Chorpani	Rishikesh	5 MLD	5.0 (approx.)	MBBR

Thus, floating populations generate a large quantity of plastic waste, and the maximum number of WWTPs are overwhelmed. Recently accumulated SS were procured from the secondary and primary sludge disposal points of the WWTPs. To prevent contamination, a sample (1 kg) of dewatered sludge (70%–80% moisture) was collected at the WWTP site and stored in glass jars with metal stoppers.

11.2.2 MP Extraction

Firstly, the moist sludge samples were kept in a hot-air oven (60°C) for 2 hours to remove the surplus moisture without affecting the MP's properties. Ten grams of uniform dried sludge samples were placed with 20 mL of 30% H_2O_2 in a beaker for 2 hours at 70°C for digestion. After a half-hour break, 10 mL of H_2O_2 was added to ensure complete organic component digestion. Wet oxidized sludge was put in a beaker with 50 mL of a NaCl solution saturated (~1.2 g/cm^3), and the mixture was left for 24 hours to create an MP layer on the surface. A conventional methodology was then used to separate the surface layer, as explained by (Zhang et al., 2020b) and the separated layer was dissolved in 50 mL Milli-Q water for an hour. The material was filtered through cellulose nitrate filter paper of size 1,000, 100, and 20 µm. In glass Petri dishes, the filter paper was oven-dried at 45°C for 3 hours. MP's weight was recorded for all the processed samples.

11.2.3 MP Characterization and Identification

MP particles were transferred to a Petri plate from filter paper containing milli-Q water and analysed with the help of a stereomicroscope for counting. MP were assessed for their particle size, particle characteristics, and colour using a binocular microscope (Nikon Eclipse-Ci) equipped with a Nikon DS–Fi2 HD camera. According to the method used by (Zhang et al., 2020a, 2020b), MP particle shapes (fibres, film, and fragments) and colours were recorded. ATR-FTIR (Shimadzu IRSpirit, Japan) analysis was used to identify the polymer types of the extracted MP samples.

11.2.4 Quality Control and Assurance

Analytical procedures employed chemicals and reagents with a purity level of 99.9% (AR grade) for all experiments. Aluminium foil was used to envelop the sludge samples, serving as a preventive measure for MP deposition through dust precipitation, and a closed fume hood was used for the extraction of MP in the laboratory. With the help of milli-Q water, the solutions and reagents were prepared to prevent contamination. After using ethanol to clean the mechanical stage, a microscopic analysis was carried out. During the experiment, tissue paper was used to clean the surface.

11.3 RESULTS AND DISCUSSION

11.3.1 Microplastic Abundance

The content of MP particles in SS samples obtained from studied WWTPs in two distinct cities within Uttarakhand, India is shown in Figure 11.2a. SS-A exhibited the highest MP content 42.4×10^3 particles/kg than SS-B 32.6×10^3 particles/kg. The load of MP was found to be in line as shown by previously reported studies on WWTP (see Table 11.1). The differences in waste management and disposal procedures between high-income countries (like Europe and the US) and middle/lower-income countries (like India and China) may be the cause of the diversity in MP loads at WWTPs.

The variation in MP load at WWTPs might be ascribed to disparities in management and disposal procedures between high-income countries such as the United States and Europe, as opposed to lower- or middle-income countries like India and China. Enhanced plastic waste management systems in the former category could be a contributing factor to the observed outcomes (Lebreton et al., 2017). The amount

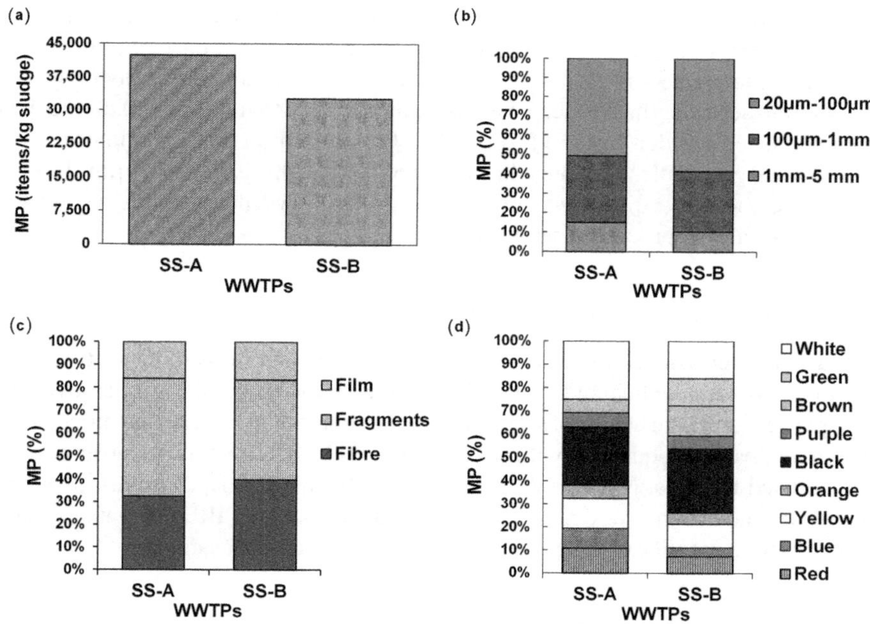

FIGURE 11.2 Sludge MP abundance: content per kg sludge (a), size distribution (b) shape distribution (c), colour abundance (d).

of MPs in sludge varies depending on the type of sludge (primary; secondary sludge; chemically stabilized sludge; dewatered sludge; activated sludge or fresh sludge) and the source inputs (number of houses in the area; industries; stormwater; landfill; municipal solid waste leachates) (Cydzik-Kwiatkowska et al., 2022; Mahon et al., 2017). The microscopic plastic debris pieces (like MP) are introduced to wastewater and urban runoff streams before being collected in sludge at WWTPs. As reported, WWTPs effectively eliminate 99.9% of MPs from wastewater. Consequently, SS could potentially serve as a prime sink of MP in the environment.

11.3.2 Microplastic Size Distribution

MPs predominantly exhibited a particle size ranging from 20 to 100 µm (53.9%) followed by 1 mm–100 µm (33.05%) and 1–5 mm (13.05%) sized MPs. Figure 11.2b shows the MP-sized distribution in SS from studied WWTPs. According to Ren et al. (2020), smaller MP particles are harmful to ecological and biological systems since they linger longer in the environment. Previous studies have also shown MPs in SS within specific size ranges: 100–500 µm (54%) and 10–100 µm (12%) (Magni et al., 2019); 250 µm–1 mm (12.65%), and 125,250 µm (39.76%) (Raju et al., 2020); 106–300 µm (Lee & Kim, 2018); <250 and 250 µm–5 mm (EL Hayany et al., 2020) from different parts of the world. The principal MP typically has a diameter of one micrometre and is mostly produced by personal care products that reach WWTPs through domestic wastewater. The size of MP, from bigger fractions to extremely small fractions, is mainly determined by the source of MP in wastewater and the ease of degradation of the plastic polymer present in the surrounding environment. However, processes such as natural weathering, UV photolytic degradations, microbiological activities, wind erosions, thermal degradation, etc. cause the breakdown of bigger plastic waste substances into smaller particles (Carr et al., 2016).

11.3.3 Microplastic Colour and Shape Pattern

Based on their shapes, MP identified in SS can be categorized into three types: fibres, fragments, and films. Figure 11.2c displays microscopic images of particular observed MP particles. The percentage of fragments, fibre, and the film was 51.41%, 32.54%, and 16.05% in SS-A sludge and 43.55%, 39.89%, and 16.56% in SS-B sludge, respectively. Fragments were the dominant MP shape in SS, which seems similar to the findings of a few earlier studies (e.g., Cydzik-Kwiatkowska et al., 2022). WWTP technologies also contributed to the fragmentation of large particles in the treatment processes. Conversely, some studies have highlighted that among various shapes (film, foam, flake, and sphere) of MPs in SS, fibres tend to be the most prevalent form as reported by a few earlier studies (Li et al., 2018; Zhou et al., 2020).

MP in the SS was found to be of various colours including black, purple, blue, green, red, white, yellow, brown, and orange (Figure 11.2d). Black colour MP was the most abundant (26.13%) followed by, white (21.07%), yellow (12.0%), red (9.33%), brown (8.80%), blue (6.40%), purple (5.87%), orange (5.33%) and green (5.07%). The origin and polymer types of MP directly affect the colour variations in SS at WWTP.

Meanwhile, the MP colour is also impacted by the weathering of environmental plastic waste (Pflugmacher et al., 2021). For instance, light colour (white and yellow) may occur as a result of plastic materials fading or wearing away, as well as hydraulic influence from the movement of MP in sewage systems and WWTP operations (Galafassi et al., 2019). The significantly high proportion of black-coloured MP in studied SS from WWTPs might be attributed to the sources, and primarily black polyethylene sheets are commonly used in agriculture and houses for mulching and construction works (Li et al., 2020). Moreover, black polyethylene carry bags are commonly used by shopkeepers for packaging vegetables/food and other grocery items in this area. According to previous studies, synthetic fibres (polystyrene), moulded furniture and household utensils (plastic nets, package food rappers, plastic substances), cosmetics, facewash scrubbers, microbeads in toothpaste, toothbrush bristles, kitchen scrubbers, and scrubbers, could be prime source of coloured-MPs in SS. Sometimes the colour misinterpreted the source and types of MP (Zhang et al., 2020a), therefore; for additional validation of the MP source, an analysis of ATR-FTIR spectra should be conducted.

11.3.4 Microplastic Polymer Characterization

ATR-FTIR spectra analysis was employed to discern the polymer types of MPs in SS collected from different WWTPs. The outcomes revealed the existence of six major polymer types and polymer groups in SS, as depicted in Figure 11.3. The content of MP polymer was found to be in order: polystyrene (PS) (36.36%) > polyethylene (PE) (27.27%) > polycarbonate (PC) (18.18%) > polyethylene terephthalate (PET) (9.09%) and nylon (9.09%) in the SS-A while polycarbonate (PC) (40%) > nylon (30%) > polyvinyl chloride (PVC) (20%) and polyethylene (PE) (10%) in the SS-B sample. Significant differences in MP polymer types were identified in both WWTPs, which may be attributed to a variety of factors such as variations in local plastic uses, waste disposal practices, mixing of urban runoff in the urban sewage line, industrial sources, open waste littering, and so on (Cydzik-Kwiatkowska et al., 2022). Figure 11.4 shows microscopic images of specific selected MP particles. Our results of polymer analysis showed consistency with previously reported results on MP contents in SS from WWTPs (see Carr et al., 2016; Hatinoğlu & Sanin, 2021).

The MP polymer types found in SS reflect the consumer behaviour, methods of waste disposal, and capabilities related to the collection of wastewater and treatment specific to the area. Uttarakhand is known for pleasant weather and adventure tourism and apart from being a hill state Uttarakhand is also known for religious tourism. Each year millions of people visit Rishikesh and adjoined Haridwar and Dehradun city. Such influx of a transient population during weekends, holidays, and festivals from across India. Consequently, a significant concern in the region pertains to the widespread open dumping of polyethylene, snack wrappers, and food packaging materials. Large plastic fragments discarded in public spaces persist for extended periods without being collected, thereby adding to the cumulative levels of MPs in the surrounding area. Over time, these fragments find their way to wastewater treatment plants through runoff, especially during monsoon seasons, and ultimately, SS from WWTP appeared as the prime sink of these MP particles.

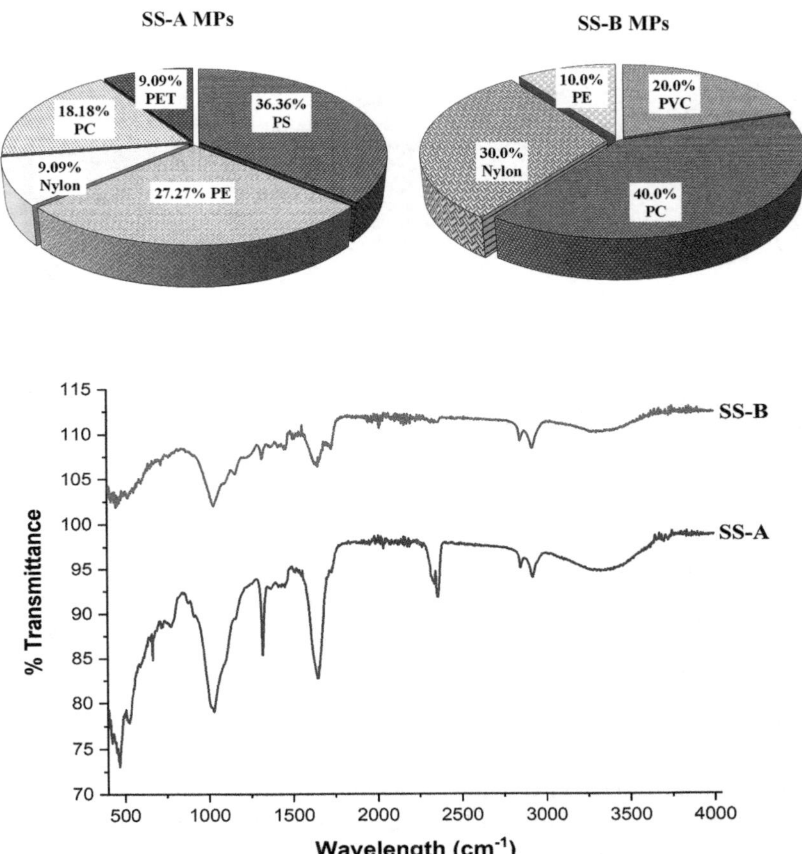

FIGURE 11.3 MP polymer types identified from SS collected from both WWTPs and FTIR spectra.

11.4 CONCLUSIONS AND FUTURE PERSPECTIVES

The purpose of this chapter was to comprehensively investigate the load and characterization of MPs in SS from two cities of Uttarakhand across the Ganga River basin which is considered to be one of the most important eco-sensitive zones in India. The results of this study revealed a significantly greater MP concentration in sludge with varying particles, various forms, and chemical origin comparable to previous studies from other parts of the world. Polystyrene and polycarbonate were identified as the predominant polymers contained in SS, mainly contributed by human sources as confirmed by ATR-FTIR result analysis. The MP abundance and its characterization are contingent on various factors such as population density, local development, habits of human beings, and treatment units in WWTPs. The dominant particle size of MPs was 20–100 μm. This study is a preliminary investigation on MP load in

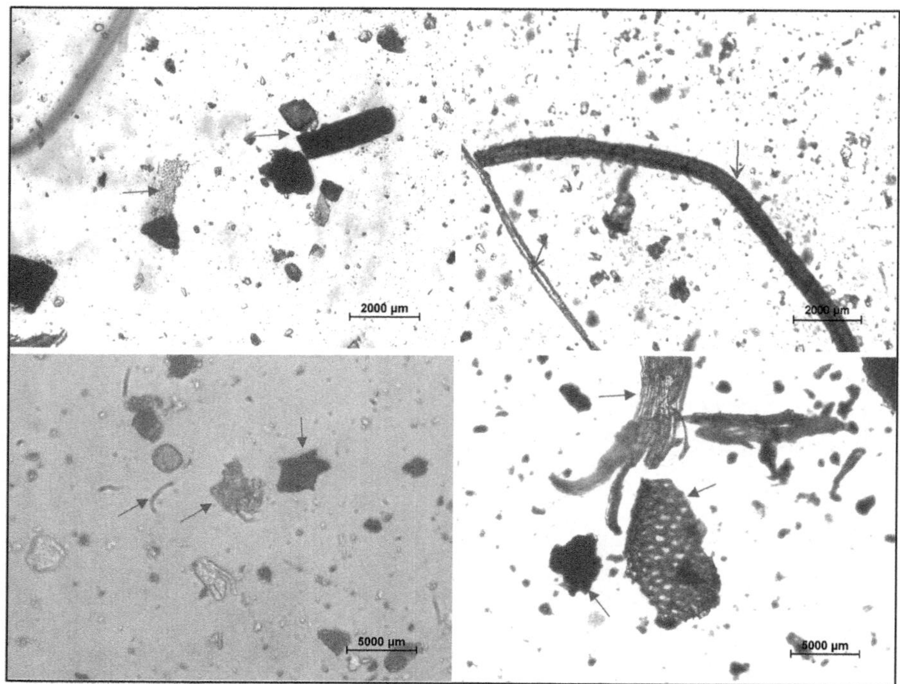

FIGURE 11.4 Microscopic photographs of MP segregated from sludge (resolution 10×).

WWTPs in upper Ganga River cities, and future research might focus on the seasonal variations of microplastic in the wastewater treatment plants and its impact on the availability of other harmful substances within the sludges and the role of WWTP technology in removing the MP at different treatment processes. The possible remediation of MP from sludge may be taken as a topic for future studies to ensure the safety of crops and soils while using such contaminated sludge as bio manure for agriculture practices.

REFERENCES

Abbasi, S., Keshavarzi, B., Moore, F., Turner, A., Kelly, F. J., Dominguez, A. O., & Jaafarzadeh, N. (2019). Distribution and potential health impacts of microplastics and microrubbers in air and street dusts from Asaluyeh County, Iran. *Environmental Pollution*, *244*, 153–164. https://doi.org/10.1016/J.ENVPOL.2018.10.039

Auta, H. S., Emenike, C. U., & Fauziah, S. H. (2017). Distribution and importance of microplastics in the marine environment: A review of the sources, fate, effects, and potential solutions. *Environment International*, *102*, 165–176. https://doi.org/10.1016/J.ENVINT.2017.02.013

Brandsma, S. H., Nijssen, P., Van Velzen, M. J. M., & Leslie, H. A. (2013) Microplastics in river suspended particulate matter and sewage treatment plants. Report R14/02, Institute for environmental studies. University Amsterdam, Netherland. Available online https://puc.overheid.nl/rijkswaterstaat/doc/PUC_147662_31.

Carr, S. A., Liu, J., & Tesoro, A. G. (2016). Transport and fate of microplastic particles in wastewater treatment plants. *Water Research*, *91*, 174–182. https://doi.org/10.1016/J.WATRES.2016.01.002.

Cydzik-Kwiatkowska, A., Milojevic, N., & Jachimowicz, P. (2022). The fate of microplastic in sludge management systems. *Science of the Total Environment*, *848*, 157466. https://doi.org/10.1016/J.SCITOTENV.2022.157466.

Dris, R., Gasperi, J., Saad, M., Mirande, C., & Tassin, B. (2016). Synthetic fibers in atmospheric fallout: A source of microplastics in the environment? *Marine Pollution Bulletin*, *104*(1–2), 290–293. https://doi.org/10.1016/J.MARPOLBUL.2016.01.006.

Edo, C., González-Pleiter, M., Leganés, F., Fernández-Piñas, F., & Rosal, R. (2020). Fate of microplastics in wastewater treatment plants and their environmental dispersion with effluent and sludge. *Environmental Pollution*, *259*, 113837. https://doi.org/10.1016/J.ENVPOL.2019.113837.

EL Hayany, B., EL Fels, L., Quénéa, K., Dignac, M. F., Rumpel, C., Gupta, V. K., & Hafidi, M. (2020). Microplastics from lagooning sludge to composts as revealed by fluorescent staining- image analysis, Raman spectroscopy and pyrolysis-GC/MS. *Journal of Environmental Management*, *275*, 111249. https://doi.org/10.1016/J.JENVMAN.2020.111249.

Fischer, E. K., Paglialonga, L., Czech, E., & Tamminga, M. (2016). Microplastic pollution in lakes and lake shoreline sediments – A case study on Lake Bolsena and Lake Chiusi (central Italy). *Environmental Pollution*, *213*, 648–657. https://doi.org/10.1016/J.ENVPOL.2016.03.012.

Galafassi, S., Nizzetto, L., & Volta, P. (2019). Plastic sources: A survey across scientific and grey literature for their inventory and relative contribution to microplastics pollution in natural environments, with an emphasis on surface water. *Science of the Total Environment*, *693*, 133499. https://doi.org/10.1016/J.SCITOTENV.2019.07.305.

Gatidou, G., Arvaniti, O. S., & Stasinakis, A. S. (2019). Review on the occurrence and fate of microplastics in sewage treatment plants. *Journal of Hazardous Materials*, *367*, 504–512. https://doi.org/10.1016/J.JHAZMAT.2018.12.081.

Hatinoğlu, M. D., & Sanin, F. D. (2021). Sewage sludge as a source of microplastics in the environment: A review of occurrence and fate during sludge treatment. *Journal of Environmental Management*, *295*, 113028. https://doi.org/10.1016/J.JENVMAN.2021.113028.

Hurley, R., Crossman, J., Schell, T., Rico, A., Futter, M., Vighi, M., & Nizzetto, L. (2020). Fate of microplastic particles in agricultural soil systems: Transport and accumulation processes in contrasting environments. *Environmental Pollution*, *293*, Article 118520. https://doi.org/10.1016/j.envpol.2021.118520.

Lassen, C., Hansen, S. F., Magnusson, K., Hartmann, N. B., Jensen, P. R., Nielsen, T. G., & Brinch, A. (2015). Microplastics: occurrence, effects and sources of releases to the environment in Denmark.

Lebreton, L. C. M., Van Der Zwet, J., Damsteeg, J.-W., Slat, B., Andrady, A., & Reisser, J. (2017). River plastic emissions to the world's oceans. *Nature Communications*, *8*, Article Number 15611. https://doi.org/10.1038/ncomms15611.

Lee, H., & Kim, Y. (2018). Treatment characteristics of microplastics at biological sewage treatment facilities in Korea. *Marine Pollution Bulletin*, *137*, 1–8. https://doi.org/10.1016/J.MARPOLBUL.2018.09.050.

Li, X., Chen, L., Ji, Y., Li, M., Dong, B., Qian, G., Zhou, J., & Dai, X. (2020). Effects of chemical pretreatments on microplastic extraction in sewage sludge and their physicochemical characteristics. *Water Research*, *171*. https://doi.org/10.1016/j.watres.2019.115379.

Li, X., Chen, L., Mei, Q., Dong, B., Dai, X., Ding, G., & Zeng, E. Y. (2018). Microplastics in sewage sludge from the wastewater treatment plants in China. *Water Research*, *142*, 75–85. https://doi.org/10.1016/J.WATRES.2018.05.034.

Long, Z., Pan, Z., Wang, W., Ren, J., Yu, X., Lin, L., Lin, H., Chen, H., & Jin, X. (2019). Microplastic abundance, characteristics, and removal in wastewater treatment plants in a coastal city of China. *Water Research, 155*, 255–265. https://doi.org/10.1016/J.WATRES.2019.02.028.

Magni, S., Binelli, A., Pittura, L., Avio, C. G., Della Torre, C., Parenti, C. C., Gorbi, S., & Regoli, F. (2019). The fate of microplastics in an Italian Wastewater Treatment Plant. *Science of the Total Environment, 652*, 602–610. https://doi.org/10.1016/J.SCITOTENV.2018.10.269.

Magnusson, K., & Norén, F. (2014). *Screening of microplastic particles in and down-stream a wastewater treatment plant*. https://www.divaportal.org/smash/record.jsf?pid=diva2%3A773505&dswid=-7010

Mahon, A. M., O'Connell, B., Healy, M. G., O'Connor, I., Officer, R., Nash, R., & Morrison, L. (2017). Microplastics in sewage sludge: Effects of treatment. *Environmental Science and Technology, 51*(2), 810–818. https://doi.org/10.1021/acs.est.6b04048.

Pflugmacher, S., Tallinen, S., Kim, Y. J., Kim, S., & Esterhuizen, M. (2021). Ageing affects microplastic toxicity over time: Effects of aged polycarbonate on germination, growth, and oxidative stress of Lepidium sativum. *Science of the Total Environment, 790*, 148166. https://doi.org/10.1016/J.SCITOTENV.2021.148166.

Plastic Europe. (2019). https://plasticseurope.org/knowledge-hub/plastics-the-facts-2019/.

Raju, S., Carbery, M., Kuttykattil, A., Senthirajah, K., Lundmark, A., Rogers, Z., SCB, S., Evans, G., & Palanisami, T. (2020). Improved methodology to determine the fate and transport of microplastics in a secondary wastewater treatment plant. *Water Research, 173*, 115549. https://doi.org/10.1016/J.WATRES.2020.115549.

Ren, X., Sun, Y., Wang, Z., Barceló, D., Wang, Q., Zhang, Z., & Zhang, Y. (2020). Abundance and characteristics of microplastic in sewage sludge: A case study of Yangling, Shaanxi province, China. *Case Studies in Chemical and Environmental Engineering, 2*, 100050. https://doi.org/10.1016/J.CSCEE.2020.100050.

Rolsky, C., Kelkar, V., Driver, E., & Halden, R. U. (2020). Municipal sewage sludge as a source of microplastics in the environment. *Current Opinion in Environmental Science & Health, 14*, 16–22. https://doi.org/10.1016/J.COESH.2019.12.001.

Talvitie, J., Mikola, A., Koistinen, A., & Setälä, O. (2017). Solutions to microplastic pollution – Removal of microplastics from wastewater effluent with advanced wastewater treatment technologies. *Water Research, 123*, 401–407. https://doi.org/10.1016/J.WATRES.2017.07.005.

Vivekanand, A. C., Mohapatra, S., & Tyagi, V. K. (2021). Microplastics in aquatic environment: Challenges and perspectives. *Chemosphere, 282*, 131151. https://doi.org/10.1016/J.CHEMOSPHERE.2021.131151.

Wang, W., Ge, J., Yu, X., & Li, H. (2020). Environmental fate and impacts of microplastics in soil ecosystems: Progress and perspective. *Science of the Total Environment, 708*, 134841. https://doi.org/10.1016/J.SCITOTENV.2019.134841.

Yu, Q., Hu, X., Yang, B., Zhang, G., Wang, J., & Ling, W. (2020). Distribution, abundance and risks of microplastics in the environment. *Chemosphere, 249*, 126059. https://doi.org/10.1016/J.CHEMOSPHERE.2020.126059.

Zhang, S., Han, B., Sun, Y., & Wang, F. (2020a). Microplastics influence the adsorption and desorption characteristics of Cd in an agricultural soil. *Journal of Hazardous Materials, 388*, 121775. https://doi.org/10.1016/J.JHAZMAT.2019.121775.

Zhang, X., Leng, Y., Liu, X., Huang, K., & Wang, J. (2020b). Microplastics' pollution and risk assessment in an urban river: A case study in the Yongjiang River, Nanning City, South China. *Exposure and Health, 12*(2), 141–151. https://doi.org/10.1007/s12403-018-00296-3.

Zhou, G., Wang, Q., Zhang, J., Li, Q., Wang, Y., Wang, M., & Huang, X. (2020). Distribution and characteristics of microplastics in urban waters of seven cities in the Tuojiang River basin, China. *Environmental Research*, *189*, 109893. https://doi.org/10.1016/J.ENVRES.2020.109893.

Zubris, K. A., & Richards, B. K. (2005). Synthetic fibers as an indicator of land application of sludge. *Environmental Pollution*, *138*(2), 201–211. https://doi.org/10.1016/J.ENVPOL.2005.04.013.

12 Distribution of Microplastic in Egypt Wastewater Using Aquatic Insects as Bioindicators

Azza M. Khedre, Somaia A. Ramadan, Ali Ashry, and Mohamed Alaraby

12.1 INTRODUCTION

Plastic manufacturing has expanded globally in recent years, making environmental contamination with plastics unavoidable (Abbing 2019; Rillig and Lehmann 2020). Big polymers are reduced in size into a variety of sizes because of environmental plastic degradation. Macro plastic is defined as being larger than 25 mm, meso plastic as being between 5 and 25 mm, big microplastic as being between 1 and 5 mm, and tiny microplastic as being between 1 and 1 mm (Immerschitt and Martens 2021). Microplastics with a variety of colors, forms, sizes, chemical compositions, and concentrations have been found in a number of freshwaters, terrestrial, and marine settings (Campanale et al. 2020). Anderson et al. (2016) stated that two likely routes that MPs could enter freshwater systems from terrestrial environments include runoff from urban, agricultural, and industrial areas and discharge from wastewater treatment facilities (WWTP). As a result, marine ecosystems have been identified as MP pollution's last sink (Nel et al. 2018).

MPs are categorized as primary or secondary based on their origins (Cole et al. 2011). The Primary MPs are generated in very small amounts via daily usage of toothpaste, lotions, and similar products that may either enter directly into freshwater resources or pass through an inefficient WWTP (Wagner and Lambert 2018). On the other hand, bigger plastics produce secondary MPs when they come into contact with biotic and abiotic agents like microorganisms and abiotic mechanical agents like wind, waves, and UV light. The example of secondary MPs includes fragments and films (Weinstein et al. 2016; Dawson et al. 2018). Regarding the interactions between MPs and freshwater biota, earlier research has looked at how MPs may be absorbed by freshwater animals from sediments and how freshwater invertebrates' capacity to absorb MPs is influenced by how they feed (Scherer et al. 2017). Macroinvertebrates are extremely valuable as bioindicators for evaluating MP pollutants in both the

DOI: 10.1201/9781003438793-12

water column and sediment in freshwater because of their abundance in many biological niches.

Despite being generally safe, MPs have certain properties (such as hydrophobicity), which varies depending on the kind of polymer, that allow them to adsorb hazardous materials including organic pollutants and heavy metals (Collicutt et al. 2019). Mao et al. (2021) claimed that MPs can absorb organic pollutants up to a million times more than the levels in the surrounding air. Furthermore, Wagner et al. (2014) discovered that as MPs get smaller, their adsorption capacity increases. Consequently, when consumed by aquatic life, MPs may contain incredibly hazardous materials (Velzeboer et al. 2014; Wang et al. 2017). Moreover, higher trophic-level species are more susceptible to the negative effects of MPs since they tend to retain MPs in their bodies for longer periods of time (Wagner and Lambert 2018). Prior research has demonstrated that plastic pollution can also occur in marine environments. Besseling et al. (2015) and Horton et al. (2018) have focused mostly on fish and other members of the top aquatic food webs' biota. A limited number of recent studies has also examined freshwater insects' MP intake (Lee and Kim 2018; Nel et al. 2018; Windsor et al. 2019; Khdre et al. 2023). Few details are known on the presence of MPs in the freshwater environments of Sohag Governorate's sediment, water, or aquatic insects. The investigations carried out on one of the Sohag Governorate's most polluted wastewater basins are reported in this study. This basin receives a large volume of wastewater, much of which is utilized for irrigation. When MPs are brought to the surface of agricultural soil as a result of wastewater sediment being utilized as fertilizer, a serious problem occurs. Therefore, the main goal of the current study is to (i) measure the number of MPs in the water and sediment of the existing wastewater basin. (ii) Check to see if MPs are present in the larvae of *Chironomus* sp., the bug with the highest abundance.

12.2 MATERIALS AND METHODS

12.2.1 Study Area

The samples were collected from a basin of Sohag Governorate, located at a distance of 10 with coordinates of 26°33″47″N and 31°36′11″E. This basin has a surface area of around 3.786 km^2 and an average depth of 1 m. The basin receives wastewater from a number of sources that contribute to pollution, such as sewage flow from nearby communities, human activity due to agricultural development and animal husbandry, and a nearby wastewater treatment facility. The basin is covered with a significant amount of algae.

12.2.2 Collection of Samples

Water samples were taken in July of 2021 at the vicinity of the basin's edge. To create an extensive distribution pattern of the MPs found in the basin, triplicate samples of the surface water were taken at five different horizontal places (pl, p2, p3, and p4) at a depth of about 10 cm. Each point is located around 150 m apart from the other. Subsequently, the water was preserved for future laboratory research. Using a stainless-steel spoon, sediment samples were obtained from each sampling location (Khedre et al. 2023 a, b). From the upper sediment (0–5 cm), 15 samples totaling a combined 2 kg were taken and transported back to the lab, where they were stored at 20°C until use. *Chironomus* sp. larvae were collected using a net with a mesh size of 0.5 mm from five different locations. Groups of collected larvae were rapidly stored in 70% alcohol in 100 mL screw-capped glass vials, as described by Nel et al. (2018), to prevent the ejection of stomach contents, which could alter MP estimate.

12.2.3 Analysis and Methods

12.2.3.1 Sediment Characteristics

For three days, the weighed sediment samples were dried at 60°C. Five subsamples of the dried sediment were used to calculate the organic content and particle size of the previous material. A mechanical sieve system incorporating several conventional sieves with differing pore diameters (4,000, 2,000, 500, 212, and 53 μm) was utilized to separate the various sediment fractions. Each sediment component was carefully collected and weighed to ascertain its quantity. Five duplicates of the dry powdery sediment, weighing around 2 g each, were used to calculate the organic content. The samples were burned for 6 hours at 600 °C in carefully weighed porcelain crucibles.

12.2.3.2 Microplastic Extraction

In order to extract MPs from water samples, Wang et al. (2019) state that 150 mL of 10% KOH was added to each sample (5 L), and the sample was left to sit for 10 hours in order to break down the organic matter. Water samples were then filtered via 0.45 μm pore size cellulose nitrate filter paper utilizing a vacuum filtration apparatus. The filter paper was placed on glass Petri plates and examined under a stereomicroscope in order to gather MPs for chemical analysis. After being put in spotless glass containers, the sediment samples were dried for 48 hours in a 60°C oven. Each sediment sample was dried and then placed in a clean 1 L beaker at a weight of 100 g. A density separation method was used to isolate MPs from denser natural particles. The freshly prepared NaCl/NaI (0.5 g/cm^3) hypersaline solution was used to dilute the sediment samples until the beaker was about half full (all solutions were filtered before use to prevent contamination) (Coppock et al. 2017). For two days, the beakers were shaken at 200 rpm using an OS-2000 open-air dual-action shaker (JEIOTECH, Korea) in order to extract any MPs (GESAMP, 2019). To remove any sediment and enable the MPs to float, the floating supernatants were moved to a second beaker and left for a full day. We repeat the earlier steps to make sure all the MP is taken out of the sediment.

Fourth instar larvae of *Chironomus* sp. were adequately cleansed to remove any unwanted items and then divided into 10 groups of 20 individuals. The weight of each was determined with the use of analytical balances. The larvae from each group were put into glass beakers, and 20 mL of H_2O_2 (35% V/V), which had been filtered beforehand, was added. The mixture was then heated in a microwave for two minutes (Windsor et al. 2019). The supernatant was dried in Petri dishes with covers in an oven at 40°C after being vacuum filtered through filter paper (0.45 μm). The MPs that were extracted were saved for later examination.

12.2.3.3 Microplastic Identification and Characterization

Using a dissecting microscope fitted with a digital camera, each MP was visually identified (Carl Zeiss Suzhou Co.). More precise confirmation of MP identification was achieved with a heated needle (Windsor et al. 2019). After that, MPs were numbered, their shape and color were determined, and at last, they were photographed. Each MP's diameter and length were measured using the Image J software. Fourier transform infrared spectroscopy (FTIR; Alpha Bruker, Germany) was used to examine the composition of MP particles. Differently colored and shaped MP particles from various samples were chosen for testing at a resolution of $2\,cm^{-1}$ between 375 and $7,500\,cm^{-1}$. With the help of the OPUS software (Bruker Optics GmbH), the data was altered. Comparing the observed spectra to the reference spectra (Primpke et al. 2018) allowed researchers to identify the type of polymer that was present. Origin PRO8 software (origin lab) was used to create the spectra figures for the reference data and the current spectra data.

12.2.4 Precautions

During the MP estimate, cotton jackets and nitrile gloves were used to prevent contamination from external MPs. Additionally, distilled water was used on a regular basis to clean the working environment and all used tools. Samples were stored in Petri plates made of coated glass. To check for any exogenous MP contamination during the analytical process, blank samples were created using the same pretreatment techniques used on the field samples (Yuan et al. 2019).

12.2.5 Statistical Analysis

Microsoft Excel was used to perform the statistical analysis (mean and standard deviation). Additionally, the IBM SPSS 21.0 application was used to do one-way ANOVA, the two-tailed test, and the paired t-test on MP-related data.

12.3 RESULTS AND DISCUSSION

12.3.1 Abundance of Microplastics in the Wastewater Basin

In Upper Egypt, one of the closed and contaminated wastewater basins was the subject of research to monitor MPs. The average amount of MPs per kilogram of dry sediment was 385 ± 38 particles/kg. However, there were 2.05 ± 0.53 particles/L on average abundant MPs in water. In comparison to earlier research, the MP level in the sediment was often rather high. This might be explained by the high organic content ($18\% \pm 1.3\%$) and fine sediment form (sewage sediment) that were found in the current basin. These characteristics may enable the sediment to bond with MPs and act as a trap for them (Anderson et al. 2016). It could take more energy to unbind this binding of MPs (Dikareva and Simon 2019; Dahms et al. 2020). Some features of the basin, including the extremely sluggish water flow that can be deemed to be stagnant water, may offer additional support for the exceptionally high number of MPs at this location. These findings support those of Nel and Froneman (2015), who discovered that lower river flow—which was linked to the highest MP abundance in sediment—was one of the most important environmental variables influencing MP distribution in the aquatic system. The installation of MPs increases with decreasing water velocity because of gravity, biofouling, and natural chemical adsorption (Mao et al. 2020). There are few MPs in the surface water of this watershed because MPs linked to sediment are hard to revive into the water (Mao et al. 2020). Furthermore, through long-term degradation and accumulation, the contained habitat of the present basin, which has a point source of MPs, may gradually increase the abundance of MPs (Eerkes-Medrano et al. 2015). Dahms et al. (2020) discovered that an urban African stream had an average MP concentration of 166.8 particles/kg dw, which is different from the previous study. Furthermore, of all the freshwater systems studied, the Rhine River had the highest concentration of MPs (4,000 particles/kg dw), according to data released by Redondo-Hasselerharm et al. (2018). There may be a sizable difference in MP concentrations between freshwater systems. In general, urbanization and population density are substantially correlated with MP abundance, and human activities are a significant source of MP pollution (Kataoka et al. 2019; Zhang et al. 2019; Naqash et al. 2020).

Because MPs are present in the sediment, benthic macroinvertebrates may consume them and reflect the MP loads in their bodies. *Chironomus* sp. larvae had an average MP load of 107 ± 36 particles/g wet weight and 100% occurrence across all samples. Because they can withstand oxygen deficiency at the bottom where MP levels are greater than in the water column, *Chironomus* sp. larvae are widespread and abundant in freshwater habitats with heavy pollution (Bere et al. 2016; Nhiwatiwa et al. 2017). Therefore, *Chironomus* sp. larvae are a prime candidate for testing MP contamination in various aquatic ecosystems. The current findings revealed that *Chironomus* sp. larvae, whose average MP load was 107 ± 36/g ww, play an important part in determining MP contamination in freshwater. However, *Chironomus* sp. larvae from the Ogun River in Nigeria had a very high abundance of MPs (291 ± 26/g w.w.), according to Akindele et al. (2020). These results suggested that the different amounts of plastic pollution in the various freshwater habitats could be related to the different MP concentrations in deposit feeders (Chironomus larvae).

12.3.2 Characteristics of MPs in the Basin

Different samples of water, sediment, and *Chironomus* sp. larvae shared comparable MP particle properties. Fibers, which made up 100% of all detected particles, were the only MP recognized in all samples based on their form (Figure 12.1).

The presence of fibers in all of the samples suggests that the principal source of MPs in the basin is most likely to be directly related to large amounts of sewage effluents from WWTP, which are primarily composed of fibers originating from textiles. Moreover, semi-urban regions surrounding the current basin may contribute to a greater abundance of MP fibers because effluent from cleaning clothes and plastic carpets was poured into the basin. According to Browne (2015), washing machine waste can include 100 fibers per liter of effluent. According to Napper and Thompson (2016), a significant number of synthetic fibers may be found in home effluent from washing machines. According to Luo et al. (2019), a significant factor in

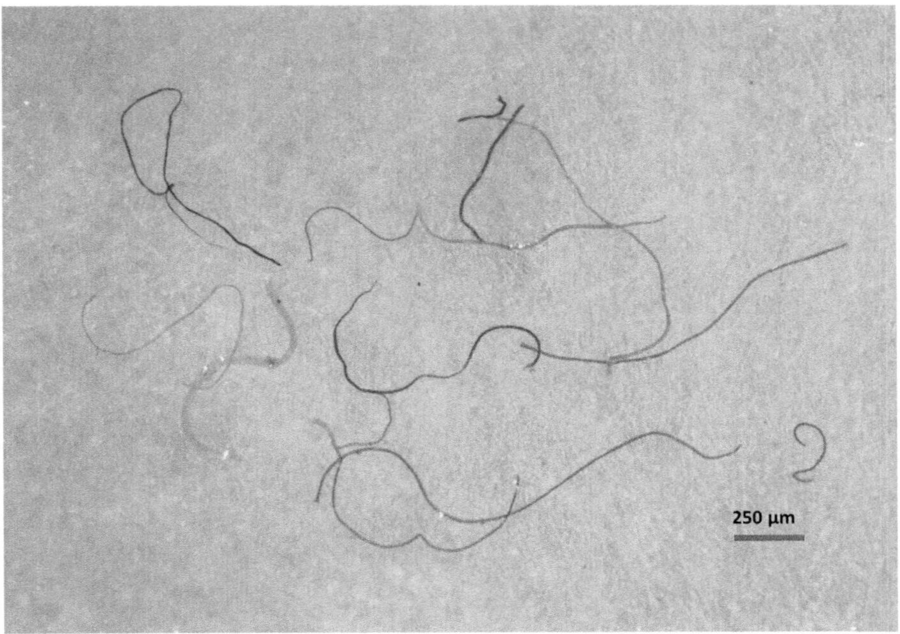

FIGURE 12.1 The photograph shows microfibers shape were collected from sediment, water, and *Chironomus* sp.

the prevalence of microfibers in aquatic systems was urban wastewater, particularly home pollution. Furthermore, agriculture was the main human activity in the region, creating a sizable number of plastic fibers (Cole et al. 2011; Horton et al. 2017). The current findings are consistent with those of earlier field freshwater studies conducted in China, where the fiber MP was found in the highest concentrations in both water and sediment in the Tiahul Lake (Su et al. 2016), the Donating Lake (41.9%, 91.9%), and the Hong Lake (44.2%, 83.9%). Additionally, Mintenig et al. (2017) investigations on treated wastewater conducted in Germany showed that synthetic fibers predominate in more than 80% of the water sample. The most common kind of MP in aquatic macroinvertebrates has also been identified as fibers, which account for more than 50% of all MPs in molluscs collected from the northern Persian Gulf (Naji et al. 2018). Additionally, microfibers made up 87% of the MPs consumed by the freshwater annelid *Tubelix tubelix* (Hurley et al. 2017). According to Akindele et al. (2019), all MPs consumed by two species of gastropods in Nigeria's Osun River were fibers. According to the current findings, all the MP fibers in the samples that were gathered were colored. There were five different colors present, and there were large percentage variances between them. In earlier studies, samples of water, sediment, and macroinvertebrates showed similar results for a wide range of colored MPs (Zhao et al. 2015). The plethora of colored plastics that are extensively produced and eaten is probably to blame for the vast spectrum of colors.

Figure 12.2a–c shows the patterns of MP color distribution in sediment, water, and *Chironomus* sp. larvae. A broad range of hues, including blue, red, black, green,

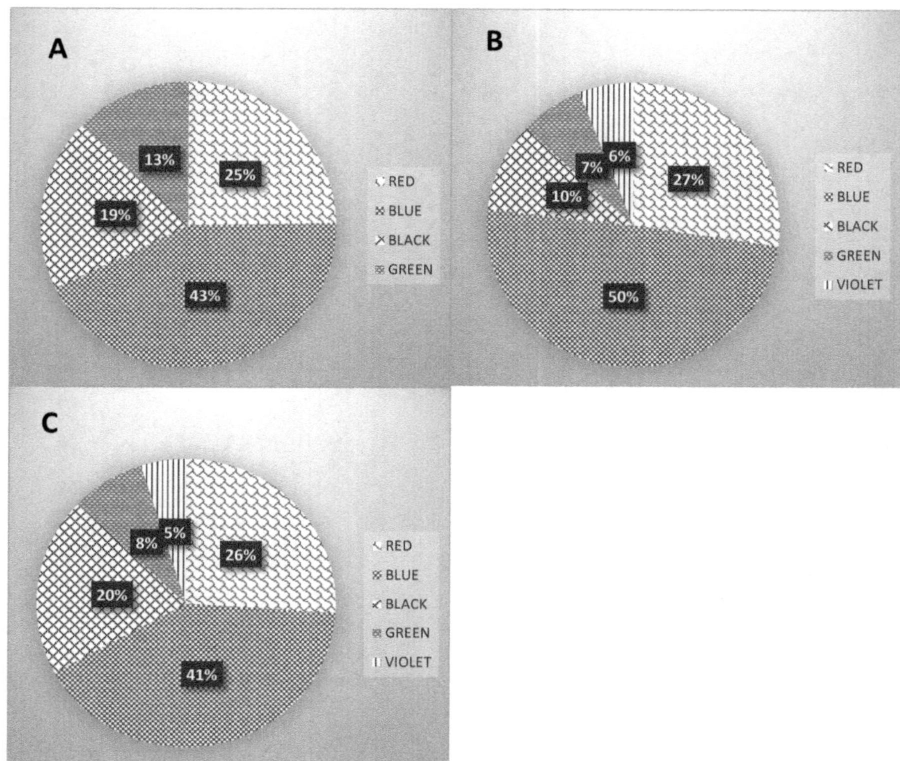

FIGURE 12.2 Relative percentages of different colored MPs were found in samples collected from (a) water, (b) sediment, and (c) *Chironomus* sp.

and violet, were discovered in microfibers. Red and blue made up 40%, 49%, and 41%, respectively, of the majority colors in all MP samples taken from water, sediment, and *Chironomus* sp. larvae. In earlier studies, samples of water, sediment, and macroinvertebrates showed similar results for a wide range of colored MPs (Zhao et al. 2015). The plethora of colored plastics that are extensively produced and eaten is probably to blame for the vast spectrum of colors. Blue is the color that is most prevalent among MPs in samples of *Chironomus* sp. larvae. This may indicate that *Chironomus* sp. larvae like this color because it may be comparable to their diet as they feed on it. According to Shaw and Day (1994), aquatic biota may consume MPs that mimic their prey in terms of color and structure.

The distribution of microfiber lengths in the various samples followed a pattern similar to that of microfiber diameters, with the longest lengths being found in sediment and water samples (Figure 12.3). The most frequent length range in sediment samples was between 1,000 and 1,500 μm, accounting for 41% of fibers found with an average length of 1.9 ± 0.5 mm, whereas in water, the average length was 1.7 ± 0.6 mm. In *Chironomus* sp. larvae, where the mean length was 0.98 ± 0.2 mm, the smaller fiber fractions were more prevalent. There was a significant difference in microfiber lengths

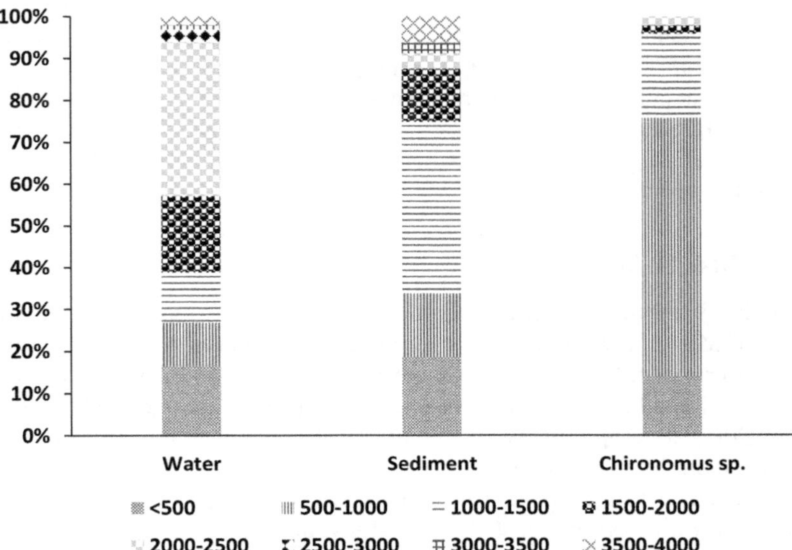

FIGURE 12.3 Relative percentages of different MP sizes were found in samples collected from water sediment and *Chironomus* sp.

between sediment and *Chironomus* sp. larvae ($p < 0.05$). It is commonly believed that the smaller the microplastic particle size, the more influence it will have on the ecosystem (Gebhardt and Forster, 2018; Li et al. 2021) since tiny MPs are more likely to accumulate in organisms. The MP size found in this investigation is categorized as "small microplastic" according to the plastic size categorization of the prior study (Gigault et al. 2021). The diameter of all the samples collected ranges from 11 to 33 µm. The small size of MPs extracted from sediment may account for the high abundance of MPs within Chironomus bodies, since the fine size is thought to be a crucial characteristic in allowing a chance for ingestion, which increases MP bioaccumulation in various organisms across food webs (Bagheri et al. 2020; Naqash et al. 2020). Furthermore, previous research (McCormick et al. 2014) found a significant abundance of microorganisms associated with MPs originating from WWTP effluent. According to Hurley et al. (2017), this may be a significant source of nutrients for aquatic biota. Additionally, it has been shown that aquatic macroinvertebrates prefer to eat detritus that has been colonized by microbes (Khedre et al. 2023c). According to a first visual assessment, certain MP fibers had organic components on them, which suggested that *Chironomus* sp. larvae may unintentionally swallow the little fibers.

Most of the MP sizes that were collected from *Chironomus* sp. larvae were significantly smaller than those that were recovered from both sediment and water. First, the larvae of *Chironomus* sp. primarily feed on small MP fibers. Secondly, MP size reduction may occur in their stomach due to internal biodegradation. This outcome can be explained by any of the two hypotheses. This finding corroborates that of Yang et al. (2014), who discovered that polyethylene MPs taken from the intestines of waxworms (*Plodia interpunctella*) were degraded by a number of bacterial species, including *Enterobacter alsuriae* and *Bacillus* sp. Based on the chemical composition of microfiber, only one polymer type—polyester—was identified among the randomly selected fibers from silt, water, and *Chironomus* sp. larvae. Polyester is frequently used in textiles, such as clothing and ropes; hence, the primary source of polyester particles in the basin may be home wastewater that is disposed of improperly.

12.4 CONCLUSIONS

This study aimed to provide a better understanding of the distribution and composition of MPs in a wastewater basin ecosystem located in a semi-urban area of Sohag Governorate, Egypt. It also set out to assess the usefulness of *Chironomus* sp. larvae as bioindicators for MP presence in freshwater, as suggested by previous studies. This was supported by our findings, which showed that *Chironomus* sp. larvae are more likely to ingest smaller microfibers from the sediment. The study also displayed that anthropogenic activities affect the type of MPs present in a certain area. Based on the results of this study, monitoring of MPs in freshwater should be carried out continually to better comprehend the impacts and implications of their presence in the ecosystem, especially since wastewater is sometimes used to irrigate agricultural land.

AUTHOR CONTRIBUTIONS

A.K. and S.R. designed this study and wrote the manuscript. A.A. and M.A. carried out field sampling along with inputs for manuscript review and data analysis.

REFERENCES

Abbing, M. R. (2019). *Plastic Soup: An Atlas of Ocean Pollution*. Island Press, Washington, DC.

Akindele, E. O., Ehlers, S. M., & Koop, J. H. (2019). First empirical study of freshwater microplastics in West Africa using gastropods from Nigeria as bioindicators. *Limnologica, 78*, 125708.

Akindele, E. O., Ehlers, S. M., & Koop, J. H. (2020). Freshwater insects of different feeding guilds ingest microplastics in two Gulf of Guinea tributaries in Nigeria. *Environmental Science and Pollution Research, 27*, 33373–33379.

Anderson, J. C., Park, B. J., & Palace, V. P. (2016). Microplastics in aquatic environments: Implications for Canadian ecosystems. *Environmental Pollution, 218*, 269–280.

Bagheri, T., Gholizadeh, M., Abarghouei, S., Zakeri, M., Hedayati, A., Rabaniha, M., … & Hafezieh, M. (2020). Microplastics distribution, abundance and composition in sediment, fishes and benthic organisms of the Gorgan Bay, Caspian sea. *Chemosphere, 257*, 127201.

Bere, T., Dalu, T., & Mwedzi, T. (2016). Detecting the impact of heavy metal contaminated sediment on benthic macroinvertebrate communities in tropical streams. *Science of the Total Environment, 572*, 147–156.

Besseling, E., Foekema, E. M., Van Franeker, J. A., Leopold, M. F., Kühn, S., Rebolledo, E. B., … & Koelmans, A. A. (2015). Microplastic in a macro filter feeder: Humpback whale Megaptera novaeangliae. *Marine Pollution Bulletin, 95*(1), 248–252.

Browne, M. A. (2015). Sources and pathways of microplastics to habitats. *Marine Anthropogenic Litter*, 229–244. https://doi.org/10.1007/978-3-319-16510-3_9.

Campanale, C., Stock, F., Massarelli, C., Kochleus, C., Bagnuolo, G., Reifferscheid, G., & Uricchio, V. F. (2020). Microplastics and their possible sources: The example of Ofanto river in southeast Italy. *Environmental Pollution*, *258*, 113284.

Cole, M., Lindeque, P., Halsband, C., & Galloway, T. S. (2011). Microplastics as contaminants in the marine environment: A review. *Marine Pollution Bulletin*, *62*(12), 2588–2597.

Collicutt, B., Juanes, F., & Dudas, S. E. (2019). Microplastics in juvenile Chinook salmon and their nearshore environments on the east coast of Vancouver Island. *Environmental Pollution*, *244*, 135–142.

Coppock, R. L., Cole, M., Lindeque, P. K., Queirós, A. M., & Galloway, T. S. (2017). A small-scale, portable method for extracting microplastics from marine sediments. *Environmental Pollution*, *230*, 829–837.

Dahms, H. T., van Rensburg, G. J., & Greenfield, R. (2020). The microplastic profile of an urban African stream. *Science of the Total Environment*, *731*, 138893.

Dawson, A. L., Kawaguchi, S., King, C. K., Townsend, K. A., King, R., Huston, W. M., & Bengtson Nash, S. M. (2018). Turning microplastics into nanoplastics through digestive fragmentation by Antarctic krill. *Nature Communications*, *9*(1), 1001.

Dikareva, N., & Simon, K. S. (2019). Microplastic pollution in streams spanning an urbanisation gradient. *Environmental Pollution*, *250*, 292–299.

Eerkes-Medrano, D., Thompson, R. C., & Aldridge, D. C. (2015). Microplastics in freshwater systems: A review of the emerging threats, identification of knowledge gaps and prioritisation of research needs. *Water Research*, *75*, 63–82.

Gebhardt, C., & Forster, S. (2018). Size-selective feeding of Arenicola marina promotes long-term burial of microplastic particles in marine sediments. *Environmental Pollution*, *242*, 1777–1786.

Gigault, J., El Hadri, H., Nguyen, B., Grassl, B., Rowenczyk, L., Tufenkji, N., ... & Wiesner, M. (2021). Nanoplastics are neither microplastics nor engineered nanoparticles. *Nature Nanotechnology*, *16*(5), 501–507.

Horton, A. A., Jürgens, M. D., Lahive, E., van Bodegom, P. M., & Vijver, M. G. (2018). The influence of exposure and physiology on microplastic ingestion by the freshwater fish Rutilus (roach) in the River Thames, UK. *Environmental Pollution*, *236*, 188–194.

Horton, A. A., Walton, A., Spurgeon, D. J., Lahive, E., & Svendsen, C. (2017). Microplastics in freshwater and terrestrial environments: Evaluating the current understanding to identify the knowledge gaps and future research priorities. *Science of the Total Environment*, *586*, 127–141.

Hurley, R. R., Woodward, J. C., & Rothwell, J. J. (2017). Ingestion of microplastics by freshwater tubifex worms. *Environmental Science & Technology*, *51*(21), 12844–12851.

Immerschitt, I., & Martens, A. (2021). Ejection, ingestion and fragmentation of mesoplastic fibres to microplastics by Anax imperator larvae (Odonata: Aeshnidae). *Odonatologica*, *49*(1–2), 57–66.

Kataoka, T., Nihei, Y., Kudou, K., & Hinata, H. (2019). Assessment of the sources and inflow processes of microplastics in the river environments of Japan. *Environmental Pollution*, *244*, 958–965.

Khdre, A. M., Ramadan, S. A., Ashry, A., & Alaraby, M. (2023). Chironomus sp. as a bioindicator for assessing microplastic contamination and the heavy metals associated with it in the sediment of wastewater in Sohag Governorate, Egypt. *Water, Air, & Soil Pollution*, *234*(3), 161.

Khedre, A. M., Ramadan, S. A., Ashry, A., & Alaraby, M. (2023a). Assessment of microplastic accumulation in aquatic insects of different feeding guilds collected from wastewater in Sohag Governorate, Egypt. *Marine and Freshwater Research*, *74*(8), 733–745.

Khedre, A. M., Ramadan, S. A., Ashry, A., & Alaraby, M. (2023b). Ingestion and egestion of microplastic by aquatic insects in Egypt wastewater. *Environmental Quality Management*, *33*, 135–145.

Khedre, A. M., Ramadan, S., Ashry, A., & Alaraby, M. (2023c). Pollution of freshwater ecosystems by microplastics: A short review on degradation, distribution, and interaction with aquatic biota. *Sohag Journal of Sciences*, *8*(3), 289–295.

Lee, H., & Kim, Y. (2018). Treatment characteristics of microplastics at biological sewage treatment facilities in Korea. *Marine Pollution Bulletin*, *137*, 1–8.

Li, J., Ouyang, Z., Liu, P., Zhao, X., Wu, R., Zhang, C., … & Guo, X. (2021). Distribution and characteristics of microplastics in the basin of Chishui River in Renhuai, China. *Science of the Total Environment*, *773*, 145591.

Luo, H., Xiang, Y., He, D., Li, Y., Zhao, Y., Wang, S., & Pan, X. (2019). Leaching behavior of fluorescent additives from microplastics and the toxicity of leachate to Chlorella vulgaris. *Science of the Total Environment*, *678*, 1–9.

Mao, R., Lang, M., Yu, X., Wu, R., Yang, X., & Guo, X. (2020). Aging mechanism of microplastics with UV irradiation and its effects on the adsorption of heavy metals. *Journal of Hazardous Materials*, *393*, 122515.

Mao, R., Song, J., Yan, P., Ouyang, Z., Wu, R., Liu, S., & Guo, X. (2021). Horizontal and vertical distribution of microplastics in the Wuliangsuhai Lake sediment, northern China. *Science of the Total Environment*, *754*, 142426.

McCormick, A., Hoellein, T. J., Mason, S. A., Schluep, J., & Kelly, J. J. (2014). Microplastic is an abundant and distinct microbial habitat in an urban river. *Environmental Science & Technology*, *48*(20), 11863–11871.

Mintenig, S. M., Int-Veen, I., Löder, M. G., Primpke, S., & Gerdts, G. (2017). Identification of microplastic in effluents of waste water treatment plants using focal plane array-based micro-Fourier-transform infrared imaging. *Water Research*, *108*, 365–372.

Naji, A., Nuri, M., & Vethaak, A. D. (2018). Microplastics contamination in molluscs from the northern part of the Persian Gulf. *Environmental Pollution*, *235*, 113–120.

Napper, I. E., & Thompson, R. C. (2016). Release of synthetic microplastic plastic fibres from domestic washing machines: Effects of fabric type and washing conditions. *Marine Pollution Bulletin*, *112*(1–2), 39–45.

Naqash, N., Prakash, S., Kapoor, D., & Singh, R. (2020). Interaction of freshwater microplastics with biota and heavy metals: A review. *Environmental Chemistry Letters*, *18*(6), 1813–1824.

Nel, H. A., & Froneman, P. W. (2015). A quantitative analysis of microplastic pollution along the south-eastern coastline of South Africa. *Marine Pollution Bulletin*, *101*(1), 274–279.

Nel, H. A., Dalu, T., & Wasserman, R. J. (2018). Sinks and sources: Assessing microplastic abundance in river sediment and deposit feeders in an Austral temperate urban river system. *Science of the Total Environment*, *612*, 950–956.

Nhiwatiwa, T., Brendonck, L., & Dalu, T. (2017). Understanding factors structuring zooplankton and macroinvertebrate assemblages in ephemeral pans. *Limnologica*, *64*, 11–19.

Primpke, S., Wirth, M., Lorenz, C., & Gerdts, G. (2018). Reference database design for the automated analysis of microplastic samples based on Fourier transform infrared (FTIR) spectroscopy. *Analytical and Bioanalytical Chemistry*, *410*, 5131–5141.

Redondo-Hasselerharm, P. E., Falahudin, D., Peeters, E. T., & Koelmans, A. A. (2018). Microplastic effect thresholds for freshwater benthic macroinvertebrates. *Environmental Science & Technology*, *52*(4), 2278–2286.

Rillig, M. C., & Lehmann, A. (2020). Microplastic in terrestrial ecosystems. *Science*, *368*(6498), 1430–1431.

Scherer, C., Brennholt, N., Reifferscheid, G., & Wagner, M. (2017). Feeding type and development drive the ingestion of microplastics by freshwater invertebrates. *Scientific Reports*, *7*(1), 17006.

Shaw, D. G., & Day, R. H. (1994). Colour- and form-dependent loss of plastic micro-debris from the North Pacific Ocean. *Marine Pollution Bulletin*, *28*(1), 39–43.

Su, L., Xue, Y., Li, L., Yang, D., Kolandhasamy, P., Li, D., & Shi, H. (2016). Microplastics in taihu lake, China. *Environmental Pollution*, *216*, 711–719.

Velzeboer, I., Kwadijk, C. J. A. F., & Koelmans, A. A. (2014). Strong sorption of PCBs to nanoplastics, microplastics, carbon nanotubes, and fullerenes. *Environmental Science & Technology*, *48*(9), 4869–4876.

Wagner, M., & Lambert, S. (2018). Microplastics are contaminants of emerging concern in freshwater environments: An overview. In Martin Wagner Scott Lambert (ed.) *Freshwater Microplastics: Emerging Environmental Contaminants?* (p. 303). Springer, Cham.

Wagner, M., Scherer, C., Alvarez-Muñoz, D., Brennholt, N., Bourrain, X., Buchinger, S., … & Reifferscheid, G. (2014). Microplastics in freshwater ecosystems: What we know and what we need to know. *Environmental Sciences Europe*, *26*(1), 1–9.

Wang, J., Liu, X., Li, Y., Powell, T., Wang, X., Wang, G., & Zhang, P. (2019). Microplastics as contaminants in the soil environment: A mini-review. *Science of the Total Environment*, *691*, 848–857.

Wang, W., Ndungu, A. W., Li, Z., & Wang, J. (2017). Microplastics pollution in inland freshwaters of China: A case study in urban surface waters of Wuhan, China. *Science of the Total Environment*, *575*, 1369–1374.

Weinstein, J. E., Crocker, B. K., & Gray, A. D. (2016). From macroplastic to microplastic: Degradation of high-density polyethylene, polypropylene, and polystyrene in a salt marsh habitat. *Environmental Toxicology and Chemistry*, *35*(7), 1632–1640.

Windsor, F. M., Tilley, R. M., Tyler, C. R., & Ormerod, S. J. (2019). Microplastic ingestion by riverine macroinvertebrates. *Science of the Total Environment*, *646*, 68–74.

Yang, J., Yang, Y., Wu, W. M., Zhao, J., & Jiang, L. (2014). Evidence of polyethylene biodegradation by bacterial strains from the guts of plastic-eating waxworms. *Environmental Science & Technology*, *48*(23), 13776–13784.

Yuan, W., Liu, X., Wang, W., Di, M., & Wang, J. (2019). Microplastic abundance, distribution and composition in water, sediments, and wild fish from Poyang Lake, China. *Ecotoxicology and Environmental Safety*, *170*, 180–187.

Zhang, J., Zhang, C., Deng, Y., Wang, R., Ma, E., Wang, J., … & Zhou, Y. (2019). Microplastics in the surface water of small-scale estuaries in Shanghai. *Marine Pollution Bulletin*, *149*, 110569.

Zhao, S., Zhu, L., & Li, D. (2015). Microplastic in three urban estuaries, China. *Environmental Pollution*, *206*, 597–604.

13 Recent Trends in Microplastic Detection based on Machine Learning and Artificial Intelligence

D. A. N. D. Daranagama and D. L. C. P. Liyanage

13.1 INTRODUCTION

Microplastics (MPs) are small plastic pieces of synthetic polymers with sizes below 5 mm and are highly persistent contaminants created a major global environmental issue in recent decades due to their toxic effects on organisms or ecosystems (Cole et al., 2011). Primarily, MPs are raw materials from some consumer products such as cosmetics and detergents. Secondary MPs that are generated via mechanical crashes, biodegradation, and photo-oxidation of primary MPs (Prata et al., 2019). To the end, currently MPs pollute all environmental systems (water, soil, air & biota). Not only that, MPs also entered organisms including humans directly through water and air and indirectly through the food chains. Once MPs enter the human body, it impacts toxicity and causes digestive problems, infections, oxidative stress and damage human cells. After accumulation MPs in the human body, they have negative health effects on inflammatory responses, metabolic disorders, nutrient absorption, and reproduction (Stapleton, 2021). There are traditional MPs identification methods such as microscopic methods and spectroscopic methods.

13.1.1 Microscopic Identification

Visual identification is one of the most common and straightforward methods used for MPs detection. It involves visually inspecting a sample under a microscope or magnifying glass to identify and quantify the MPs present in the sample. To perform visual identification, a sample is typically collected from the environment, such as seawater, sediment, or soil. The sample is then filtered or centrifuged to concentrate the MPs, and the resulting pellet is examined under a microscope. MPs can be identified based on their physical characteristics, such as size, shape, and color (Mariano et al., 2021). Typically, a sample is examined by a trained expert who visually counts and identifies the MPs present in the sample.

DOI: 10.1201/9781003438793-13

13.1.1.1 Stereo Microscopy

Stereo microscopy is a commonly used technique for the detection of MPs in environmental samples. This type of microscopy provides a 3D view of the sample, allowing for the identification and quantification of MPs based on their physical characteristics, such as size, shape, and color. Stereo microscopy is particularly useful for larger MPs that can be observed with the naked eye or under low magnification (Shim et al., 2017).

13.1.1.2 Scanning Electron Microscopy (SEM)

SEM is another powerful microscopy technique used for MP detection. SEM provides high-resolution images of the sample surface, allowing for the identification and characterization of MPs based on their morphology and surface features (Fries et al., 2013). SEM is particularly useful for smaller MPs, such as microfibers and microbeads, which can be difficult to observe with other microscopy techniques.

13.1.1.3 Confocal Microscopy

Confocal microscopy allows for the 3D imaging of a sample with high resolution and can be used to identify and quantify MPs based on their fluorescence properties. Fluorescent molecules such as Nile red used to distinguish MPs according to their hydrophobicity (Meyers et al., 2022). Fluorescent dye Nile blue or 4-chloro-7-nitro-1,2,3-benzoxadiazole was used to detect polystyrene (PS) particles in plant tissues by using a confocal laser scanning microscope and characterized MPs by their size ranging from sub-micrometer to micrometer-sized (Li et al., 2020). However, Nile red dyes were applicable to some biological samples. Therefore, the method was developed after staining with a blend of Calcofluor white and Evans blue dyes in addition to Nile red; MPs in ground arthropod biomass were detected under laser scanning confocal microscopy (Maxwell et al., 2020).

13.1.1.4 Atomic Force Microscopy (AFM)

Atomic force microscopy (AFM) can be used as a powerful tool for the identification and characterization of MPs. AFM allows for high-resolution imaging and surface analysis, which can provide valuable information about the morphology, size, and composition of MP particles. AFM can provide 3D surface profiles with nanoscale resolution (Akhatova et al., 2022) and measure the dimensions of the particles using the AFM software analysis tools.

13.1.2 Analysis Methods

13.1.2.1 Fourier Transform Infrared (FTIR) Spectroscopy

Fourier Transform Infrared (FTIR) spectroscopy is a powerful analytical technique that has been used for the detection and identification of MPs in environmental samples. FTIR spectroscopy measures the absorption or transmission of infrared light by a sample, providing information about the molecular composition and structure of the sample (Xu et al., 2019).

13.1.2.2 Raman Spectroscopy

Raman spectroscopy is another analytical technique that has been used for the detection and identification of MPs in environmental samples. Like FTIR spectroscopy, Raman spectroscopy provides information about the molecular composition and structure of a sample based on its interaction with light (Xu et al., 2019). In Raman spectroscopy, a sample is exposed to a monochromatic laser beam, and the scattered light is collected and analyzed. The resulting Raman spectrum provides information about the vibrational modes of the sample, which can be used to identify the types of chemical bonds present in the sample and to distinguish between different types of plastics based on their chemical composition (Lenz et al., 2015).

13.1.2.3 Thermal Analysis

Thermal analysis measures changes in the physical and chemical properties of polymers depending on their thermal stability. There are several thermal analysis techniques that can be used for MP identification. The two commonly employed methods are Differential Scanning Calorimetry (DSC) and Thermogravimetric Analysis (TGA). DSC measures the heat flow associated with phase transitions and thermal events (Bitter & Lackner, 2021), while TGA measures the weight change of a sample temperature (Mansa & Zou, 2021).

Overall, microscopy techniques and analysis methods are essential tools for the detection and characterization of MPs in various environmental and living samples. The choice of microscopy technique depends on the size and type of MPs present in the sample and the information required for further analysis. The development of automated microscopy techniques and image analysis methods has the potential to improve the speed and accuracy of MP detection and further our understanding of the impacts of MP pollution on the environment and living beings.

13.2 MACHINE LEARNING (ML) AND ARTIFICIAL INTELLIGENCE (AI) IN MP DETECTION

13.2.1 Machine Learning (ML) Algorithms for MP Detection

The current techniques used to identify MPs have shown notable drawbacks, including inadequate precision, time-consuming imaging processes, and restricted analysis of particle sizes. Consequently, there is a growing interest in exploring novel approaches for MP identification, focusing on advancements in research. Notably, ML algorithms have emerged as immensely valuable tools in this field, exhibiting their effectiveness in recent years (Hufnagl et al., 2021). In a review paper published by Zhang et al. (2023), significant advancements in the identification of MPs using rapid AI technology are highlighted. These advancements encompass various aspects such as automatic quantification methods, AI algorithms, and optimization techniques, ultimately leading to improved identification and understanding of MP pollution (He et al., 2016).

Over the past few years, ML algorithms have made significant strides in enhancing image recognition technology across various domains (Massarelli et al., 2021). This progress can be attributed to the advancements in automation and the widespread availability of computational resources (Yu & Hu, 2022). Researchers have leveraged remote sensing tools and object-oriented ML approaches to effectively measure and categorize plastic waste found on beaches. This innovative combination of technologies allows for a comprehensive assessment of the environmental pollution risks associated with beach plastics. Identifying MPs using ML involves training a model to recognize patterns and characteristics that distinguish MPs from other substances. ML is a data-driven approach, and having a comprehensive and sufficient database is indeed crucial for conducting global-scale or large-scale MPs research based on ML (Yu & Hu, 2022). By merging ML techniques with traditional methods such as Raman spectroscopy, the identification of MPs in intricate environments can be remarkably enhanced. This approach involves analyzing spectral data and employing correlation algorithms that leverage image recognition. Each ML algorithm possesses unique characteristics, often relying on the remarkable learning capabilities of deep neural networks, thereby improving quantification and accuracy rates (Yu & Hu, 2022). Notably, this integration offers additional advantages such as cost and time reduction in the analysis process, presenting promising prospects for MPs identification.

13.2.2 Features of MPs

As explained above, MPs can be classified based on their chemical compositions. However, this classification can be costly and may vary from region to region (Lusher et al., 2014). Therefore, ML primarily utilizes physical morphology for classification. The following features are specifically considered when analyzing digital photo images for MPs (Chaczko et al., 2018).

- Size: According to the definition, MPs are smaller than 5 mm. Therefore, if a camera-based solution is used, the detection algorithm should focus on objects measuring 5 mm and below.
- Shape: MPs can exhibit a wide variety of shapes, including both standard geometric shapes and irregular forms. The consideration of shape as a feature may vary and needs to be assessed during experimental runs. MPs can primarily be categorized into five main shapes: fiber/line, bead/pellet, fragment, film, and foam (Lusher et al., 2014; Wagner et al., 2014).
- Color: MPs can be found in different colors, which may either be similar to or distinct from other objects and particles present in oceans. Some plastic particles can be transparent or colorless and cannot be observed in the visual spectrum. However, in certain cases, MPs can have unique and easily observable colors.
- Texture: The texture of MPs can be a crucial characteristic for developing a detection scheme. Plastics manufactured through specific molding and machining processes may exhibit a consistent surface finish pattern. This pattern can be measured using texture features extracted from digital images.
- Chemical Composition: Chemical analysis techniques, such as gas chromatography-mass spectrometry (GC-MS). Features derived from chemical analysis data can include the presence or abundance of specific chemicals

or additives associated with MPs. GC-MS is indeed capable of identifying the specific polymer type present in MPs and can distinguish between different polymers such as polyethylene (PE), polypropylene (PP), polyvinyl chloride (PVC), and PS based on their unique chemical signatures. This capability is crucial for understanding the composition of MPs in environmental samples.

The selection of features in data analysis and ML depends on various factors, including the available data, the nature of the problem, and the goals of the analysis or model. It is common to use combinations of different types of features to capture multiple aspects of MPs and achieve more accurate identification or classification results. Feature engineering and selection techniques, such as dimensionality reduction methods or feature importance analysis, can help in determining the most informative and relevant features for MP identification tasks. The shapes of MPs can vary significantly, even if they have the same chemical composition, due to variations in manufacturing and degradation processes. Additionally, distinguishing between materials with similar shapes but different chemical compositions is challenging without chemical characterization (Free et al., 2014). However, it is still possible to classify objects in SEM images based on their shapes and structures with high accuracy (Aversa et al., 2018). To acquire, select, and classify appropriate feature subsets, the implementation of intelligent systems with ML algorithms is necessary. These technologies will enable the system to efficiently gather data, identify relevant features, and categorize them accurately. The system can capture high-quality data from the environment, while intelligent algorithms will analyze the data and determine the most relevant subsets of features for further processing. This combination of intelligent systems with ML algorithms will enhance the effectiveness and efficiency of the feature selection and classification processes.

13.2.3 ML Techniques

The development of ML models involves following several important steps, such as data acquisition, data preprocessing, ML model selection and evaluating results based on unknown data to enhance the accuracy of the models. Finally, the trained model is implemented to fulfill the requirements. Based on the features explained above, ML can be used in the analysis process. Mainly, two data sources of visual identification, commonly employed in both traditional ML and advanced deep learning techniques, are described below.

Normal Camera Image-Based MP Detection

- Normal camera images can be utilized for MP identification using visual analysis techniques in computer vision (Lorenzo-Navarro et al., 2021).
- Applying image segmentation algorithms to separate MPs from the background.
- Feature extraction from images using methods such as texture analysis and shape descriptors.

Spectroscopic analysis for MP detection

- Using spectroscopic techniques like FTIR or Raman spectroscopy for MP identification (Primpke et al., 2017; Wright et al., 2019).
- Preprocessing and analyzing spectral data to extract relevant features.

All ML models, including deep learning, used for MPs identification can be classified into two approaches: supervised learning and unsupervised learning. These two approaches have unique advantages for analyzing the given data based on the requirements.

1. A supervised learning algorithm, such as a convolutional neural network (CNN), can be trained on a dataset of labeled MP images. Each image is labeled with the corresponding type of MP (e.g., fiber, fragment). The CNN learns to extract features from the images and can classify new MP images based on their visual characteristics (Krogh & Vedelsby, 1994).
2. Unsupervised learning algorithms, on the other hand, operate on unlabeled datasets without predefined class labels. These algorithms aim to discover patterns, structures, or relationships within the data without any prior knowledge. In the context of MPs identification, unsupervised learning algorithms can be used when the data is unlabeled or when the objective is to explore the underlying structure of the MP dataset. An unsupervised learning algorithm, such as k-means clustering, can be applied to a dataset of MP features extracted from various sources (e.g., images, spectra). The algorithm groups the MP samples into clusters based on their similarity in feature space. This clustering can help identify distinct groups of MP with similar characteristics.

In practice, a combination of supervised and unsupervised learning techniques can be used. For instance, unsupervised learning algorithms can be applied to explore and preprocess the data, while supervised learning algorithms can be employed to build accurate classifiers or identification models based on the labeled data (Suárez-Araujo et al., 2016). The workflow for MP detection using ML encompasses data collection, data preprocessing, feature extraction, model selection, model training, model evaluation, and model deployment. Figure 13.1 shows the most essential steps for developing an effective MP detection system using ML techniques.

13.2.3.1 Image Classification

For MP images, visual features extracted from images play a crucial role. These features can include shape descriptors (e.g., aspect ratio, circularity), color features (e.g., RGB values, color histograms), texture features (e.g., Haralick features, local binary patterns), and size-related features (e.g., area, perimeter). These features capture the visual characteristics of MPs, allowing ML algorithms to distinguish between different types or categories of MPs. MPs can be visually identified using microscopy or imaging techniques. ML models, such as CNNs, can be trained to classify MP images into different categories based on their shape, color, texture, or other visual

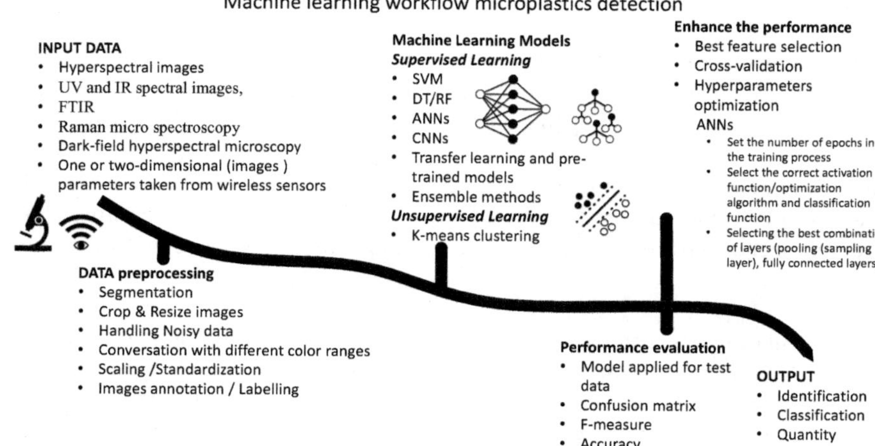

FIGURE 13.1 ML workflow for MP detection.

features. The models learn from a labeled dataset of MP images and can then classify new images with a high degree of accuracy.

CNN consists of three essential layers: convolution, pooling, and fully connected (FC) layers. The convolution and pooling layers play a crucial role in extracting features, while the fully connected layer maps these extracted features to the final output, often used for tasks like classification. The depth of the network, determined by the number of layers, can enrich the level of extracted features. This increased depth can have a positive impact on classification accuracy. However, there comes a point where increasing the depth can lead to two challenges: overfitting and degradation (Yamashita et al., 2018; Monti et al., 2018). In the case of overfitting, the model becomes excessively specialized to the training data and fails to generalize well to new, unseen data. In the case of degradation, the deeper model becomes more challenging to train due to optimization difficulties.

To address these challenges, many applications utilize transfer learning, which leverages the concept of pre-trained models to solve deep learning problems (Yamashita et al., 2018). Transfer learning involves utilizing the knowledge and learned features from models trained on large datasets, thereby providing a head start and improving the performance of models in specific tasks.

Recurrent Neural Networks (RNNs) are suitable for sequential data, such as time series or textual information. Flexibility is another advantage, as it allows for adaptation to various tasks, including sequence prediction, language modeling, and sequence-to-sequence tasks. Additionally, RNNs provide some level of interpretability because they maintain hidden states that can be examined to understand how the model processes input sequences. RNNs can be used to analyze and classify MP-related data, such as concentration measurements over time or textual descriptions of MP samples. Long Short-Term Memory (LSTM) and Gated Recurrent Unit (GRU) are popular RNN variants that can capture temporal dependencies in the data. LSTMs are effective at capturing long-range dependencies in sequential data, making them suitable for tasks where understanding context over a large time window is critical. They mitigate the vanishing gradient problem, enabling more stable training on deep networks and long sequences. Additionally, GRUs have three advantages: simplicity, good generalization, and parallelization. These factors support making predictions with computational efficiency, as GRUs have fewer parameters than LSTMs, allowing for faster training and requiring less data. Generative Adversarial Networks (GANs) are composed of two neural networks: a generator and a discriminator. GANs have been employed to generate synthetic MP images, which can be used to augment limited training datasets. GANs can also be used for data augmentation or to learn the underlying distribution of MPs, enabling the generation of realistic MP samples.

Transfer learning leverages pre-trained deep learning models that were trained on large-scale datasets, such as ImageNet, and fine-tunes them for MP identification tasks. By starting with a model that has learned general image features, transfer learning enables the efficient training of MP classification models even with limited labeled MP datasets. MP identification often involves multiple modalities, such as images, spectral data, and chemical analysis. Deep learning techniques can be used to fuse information from different modalities by designing multimodal architectures that combine convolutional, recurrent, or fully connected layers to jointly process and classify data from multiple sources. It's worth mentioning that deep learning approaches often require large, labeled datasets for training, which can be a challenge in the case of MPs. Data collection and annotation efforts need to be considered to build robust deep learning models for MP identification or classification.

13.2.3.2 Spectral Analysis

MPs can also be characterized based on their spectral properties, such as absorption, reflectance, or fluorescence. ML algorithms can be trained using spectral data from different types of MPs. These models can then predict the type or composition of MPs based on their spectral signatures. Support vector machines (SVMs) are highlighted as a powerful tool for solving small-scale, nonlinear, and high-dimensional pattern recognition problems (Melgani & Bruzzone, 2004). SVM offers unique advantages in achieving high-quality results with a small sample set. It played a significant role in addressing challenges encountered in image classification before the advent of CNN, such as low computer performance and limited availability of image samples (Imavathy & Chinnadurai, 2021). SVM algorithms have proven to be highly accurate and robust across various types and sizes of MPs. For instance, a quick detection approach for soil MP contamination was presented using near-infrared (NIR) spectroscopy. MPs in soil could be detected effectively. Additionally, NIR spectroscopy in combination with SVM regression techniques enabled efficient quantification of MPs contamination in soil (Zhao et al., 2021). SVM classifiers have demonstrated their effectiveness in overcoming limitations associated with Fourier Transform Infrared (FTIR) imaging, including large data volumes and poor spectral quality. This capability allows for rapid and automated characterization of MPs.

As a summary, SVM algorithms provide a powerful solution for MP identification, offering high accuracy and robustness in diverse MP types and sizes. They enable efficient detection and quantification of MPs in various environments, such as soil, through techniques like hyperspectral imaging and NIR spectroscopy. A novel approach of SVM, known as Holographic SVM, has achieved remarkable results in the identification of MPs in heterogeneous pretreated water samples. This innovative method has demonstrated an impressive classification accuracy of over 99% (Bianco et al., 2020).

13.2.3.3 Raman Spectroscopy

Raman spectroscopy is a technique used to analyze the molecular composition of materials, including MPs. ML algorithms can be applied to Raman spectra data to classify MPs into different polymer types or identify specific additives or contaminants present in the samples. Techniques such as principal component analysis (PCA), t-SNE or deep learning models can be used to extract features from Raman spectra and perform classification. PCA is an ML technique used for feature extraction. It aims to minimize the mean square error under statistical significance and extract new features that retain important information from the original pattern classes. PCA has been widely applied in various fields, including noise reduction in images for pattern recognition. Wander et al. (2020) applied PCA to reduce the dimensionality of MP samples, allowing for visualization of particle similarity. This approach helped in identifying key information and patterns in the μ-FTIR data. Raman spectral matrices contain dispersed information throughout the dataset (Shinzawa et al., 2009), making it necessary to minimize information loss and background interference when interpreting Raman data. PCA has proven to be an effective statistical method for handling complex Raman data matrices (Halstead et al., 2018; Jolliffe and Cadima, 2016). Fang et al. (2022) utilized PCA to decode Raman spectral matrix imaging, enabling the efficient extraction and decoding of crucial information. This process improved the accuracy and interpretability of MP samples containing numerous unknown components. Overall, PCA is a valuable tool for feature extraction in MP recognition. Its application, both in combination with other methods and standalone, has demonstrated its ability to enhance accuracy and provide insights into the underlying patterns of MP samples.

The decision tree (DT) is a popular classification algorithm used to construct a set of base classifiers based on training data. Random Forest (RF) is an extension of DT that reduces overfitting by creating multiple uncorrelated DT and averaging their predictions. In recent years, DT and RF have been increasingly applied in ecological and environmental research, including the identification of MPs. The method uses RF and hyperspectral images to identify MPs in environmental samples, achieving high accuracy with limited parameters. Jie et al. (2021) proposed an intelligent classification method that combines wavelet processing, Raman spectroscopy, and DT/RF algorithms for the quick identification of MPs in seawater. RF outperformed DT in cross-validation accuracy, reaching 97.24% accuracy, setting a benchmark for Raman spectroscopy in MP identification.

13.2.3.4 Hybrid Approaches

Combining multiple data sources and techniques can enhance the accuracy of MP identification and classification. For example, integrating image analysis with spectral or chemical data can provide a more comprehensive understanding of MPs. Hybrid

TABLE 13.1
A Summary of Inputs, Techniques, Models, and the Specific Advantages Obtained from Supervised Learning

Inputs	Attributes	Feature Extraction/ Selection Method	ML Model	Output
Hyperspectral images of textile samples (Chaczko et al., 2019)	Different texture patterns	PCA	ANN with 39 layers	Classification approach – accuracy of 95%
Monitor water surface characteristics based on UV and IR spectral images using multi-camera acquisition (Chaczko et al., 2019)	Size, shape, color, texture	Based on ANNs	Comparison among KNN, SVM, RF ANNs	Classification approach, best accuracy is obtained from ANNs
Dark-field images of the single-type PS microparticles, and the mixtures of PS microparticles	Color (dark blue, dark red, and opaque)	Based on ANNs	Deep CNN- ResNet neural network model	Average accuracy of more than 85% both single and the mixtures of PS

models that combine CNNs with other ML algorithms, such as SVMs or DTs, can improve overall classification performance. Table 13.1 provides a summary of inputs, techniques, models, and the specific advantages obtained from each approach. It is important to note that the availability and quality of labeled datasets are crucial for training accurate ML models. Collecting and annotating large and diverse datasets of MP samples is an ongoing challenge in this field. Additionally, regular model validation and testing on independent datasets are essential to ensure the reliability of the ML-based MP identification systems.

13.2.4 AI-Based Technologies

ML techniques can be effectively combined with AI methods to enhance MP detection. Most of the operations involve the application of a combination of ML techniques and AI for MP detection. This is particularly evident in processes such as data collection and preprocessing, feature engineering, data fusion, customization of prediction models using deep learning, real-time monitoring, integration with environmental factors, the creation of a user-friendly automated system, continuous learning, validation, and evaluation, as well as collaboration and data sharing. AI-based MP-imaging technologies rely on a combination of pivotal components, notably smart Unmanned Aerial Vehicles (Martin et al., 2018). This method stands as a commendable, reproducible, and efficient approach for assessing beaches worldwide. This is in stark contrast to the outdated, inefficient methodologies and disparate protocols that hinder the seamless integration and comprehensive data collection required on a national scale. These three science mapping tools—CiteSpace, VOSviewer, and Scimago Graphica—serve as invaluable aids in unraveling the intellectual foundations and exploring the cutting-edge research frontiers within the realm of AI-based MP-imaging (Stopar & Barto, 2019). By leveraging these science mapping tools, researchers can gain a comprehensive understanding of the intellectual foundations, emerging trends, and collaborative networks in AI-based MP-imaging, ultimately facilitating more informed decision-making and research strategy development. The examination of plastisphere biomarkers was employed to assess their implications on microbial ecology within the emerging anthropogenic environment.

A novel method for the categorization of microbeads (MBs) by analyzing microscopic images using CNN was presented. This innovative technique was designed to classify and characterize MPs, and it yielded an impressive classification accuracy of 89% when applied to MBs found in wastewater samples (Yurtsever & Yurtsever, 2019). Introducing an innovative methodology that marries 3D coherent imaging with the power of artificial intelligence to attain precise and automated identification of MPs within filtered water specimens spanning a broad microscale spectrum. This approach shows the possibilities of detecting the MPs via environmental monitoring. The MPs counting when the range 1–5 mm count is very hard work. Therefore, an application of AI is presented with the aim of automatically counting and classifying MP with the U-Net neural network (Lorenzo-Navarro et al., 2021). There is a notable absence of a comprehensive framework that can seamlessly link microscopy observations with material properties across a wider range of parameters, especially in MP detection. Addressing this gap, we introduce “AtomAI,” an open-source software solution that serves as a bridge connecting instrument-specific Python libraries, deep learning capabilities, and simulation tools, all within a unified ecosystem. AtomAI enables the direct utilization of deep neural networks to perform precise atomic and mesoscopic image segmentation, transforming image and spectroscopy data into localized descriptors categorized by class (Ziatdinov et al., 2021). It is speeding up the detection process and reducing the need for manual labor. AI can indeed integrate data from a wide range of sensors and sources, such as imaging, spectroscopy, and environmental monitoring, to create a comprehensive overview of MPs contamination. Additionally, the availability of a spectral library enhanced by advanced AI techniques can significantly benefit analysis software by reducing the reliance on expert time, improving analysis speed, and enhancing the accuracy of results (Hufnagl et al., 2021). Combining ML techniques with AI for MPs detection creates a robust and adaptive system that can analyze diverse data sources, adapt to changing conditions, and provide valuable insights for environmental monitoring and management. To predict and explain the movement and distribution of MPs, it is essential to consider incorporating environmental factors such as weather, water currents, and temperature. One way to achieve this is by using AI algorithms. Research introduced a mechanism for detecting and imaging the global distribution of ocean MPs from space using spaceborne bistatic radar measurements of ocean surface roughness (Evans & Ruf, 2021).

13.3 CONCLUSION AND FUTURE DIRECTIONS

Identifying MPs using ML involves training a model to recognize patterns and characteristics that distinguish MPs from other substances. ML is a data-driven approach, and having a comprehensive and sufficient database is indeed crucial for conducting global-scale or large-scale MP research on ML. With the advancement of technology, the development of IoT devices is expected to incorporate various architectural requirements, including portability, availability, automation, accuracy, robustness, scalability, and cost-effectiveness (Chaczko et al., 2018). We hope in the future, the growing datasets with more and more detailed morphological information provided by the advanced microscopes could make it possible to automatically identify and quantify the MPs in environmental samples, thus helping us monitor MP pollution. Emerging technologies and approaches in MP detection, Collaborative efforts and initiatives in the field, and Ethical considerations and responsible use of AI in MP detection will be crucial for MPs detection. The sub-topics provide a comprehensive overview of various aspects related to ML and AI are Advance technology and deploying ML models on edge devices for real-time detection. Feature selection and dimensionality reduction, Techniques to select relevant features for MP detection and utilizing pre-trained models for MP detection are challenges and considerations for real-time MP detection.

REFERENCES

Akhatova, F., Ishmukhametov, I., Fakhrullina, G., & Fakhrullin, R. (2022). Nanomechanical atomic force microscopy to probe cellular MPs uptake and distribution. *International Journal Molecular Science*, *23*(2), 806. https://doi.org/10.3390/ijms23020806.

Aversa, R., Modarres, M. H., Cozzini, S., Ciancio, R., & Chiusole, A. (2018). The first annotated set of scanning electron microscopy images for nanoscience. *Science Data*, *5*(1), 1–10. https://doi.org/10.1038/sdata.2018.172.

Bianco, V., Memmolo, P., Carcagnì, P., Merola, F., Paturzo, M., Distante, C., & Ferraro, P. (2020). MP identification via holographic imaging and machine learning. *Advanced Intelligent Systems*, *2*(2), 1900153. https://doi.org/10.1002/aisy.201900153.

Bitter H., & Lackner S. (2021). Fast and easy quantification of semi-crystalline MPs in exemplary environmental matrices by differential scanning calorimetry (DSC). *Chemical Engineering Journal*, *423*, 129941. https://doi.org/10.1016/j.cej.2021.129941.

Chaczko, Z., Kale, A., Santana-Rodríguez, J. J., & Suárez-Araujo, C. P. (2018, June). Towards an IoT based system for detection and monitoring of MPs in aquatic environments. In *2018 IEEE 22nd International Conference on Intelligent Engineering Systems (INES)* (pp. 000057–000062). IEEE. https://doi.org/10.1109/INES.2018.8523957.

Chaczko, Z., Wajs-Chaczko, P., Tien, D., & Haidar, Y. (2019, July). Detection of MPs using machine learning. In *2019 International Conference on Machine Learning and Cybernetics (ICMLC)* (pp. 1–8). IEEE. https://doi.org/10.1109/ICMLC48188.2019.8949221.

Cole, M., Lindeque, P., Halsband, C., & Galloway, T. S. (2011). Microplastics as contaminants in the marine environment: A review. *Marine Pollution Bulletin*, *62*(12), 2588–2597. https://doi.org/10.1016/j.marpolbul.2011.09.025.

Evans, M. C., & Ruf, C. S. (2021). Toward the detection and imaging of ocean MPs with a spaceborne radar. *IEEE Transactions on Geoscience and Remote Sensing*, *60*, 1–9.

Fang, C., Luo, Y., Zhang, X., Zhang, H., Nolan, A., & Naidu, R. (2022). Identification and visualisation of MPs via PCA to decode Raman spectrum matrix towards imaging. *Chemosphere*, *286*, 131736. https://doi.org/10.1016/j.chemosphere.2021.131736.

Free, C. M., Jensen, O. P., Mason, S. A., Eriksen, M., Williamson, N. J., & Boldgiv, B. (2014). High-levels of MP pollution in a large, remote, mountain lake. *Marine Pollution Bulletin*, *85*(1), 156–163. https://doi.org/10.1016/j.marpolbul.2014.06.001.

Fries, E., Dekiff, J. H., Willmeyer, J., Nuelle, M. T., Ebert, M., & Remy, D. (2013). Identification of polymer types and additives in marine MP particles using pyrolysis-GC/MS and scanning electron microscopy. *Environmental Science: Processes & Impacts*, *15*(10), 1949–1956. https://doi.org/10.1039/c3em00214d.

Halstead, J. E., Smith, J. A., Carter, E. A., Lay, P. A., & Johnston, E. L. (2018). Assessment tools for MPs and natural fibres ingested by fish in an urbanised estuary. *Environmental Pollution*, *234*, 552–561. https://doi.org/10.1016/j.envpol.2017.11.085.

He, K., Zhang, X., Ren, S., & Sun, J. (2016). Identity mappings in deep residual networks. In *Computer Vision–ECCV 2016: 14th European Conference, Amsterdam, The Netherlands, October 11–14, 2016, Proceedings, Part IV 14* (pp. 630–645). Springer International Publishing. https://doi.org/10.1007/978-3-319-46493-0_38.

Hufnagl, B., Stibi, M., Martirosyan, H., Wilczek, U., Möller, J. N., Löder, M. G., ... & Lohninger, H. (2021). Computer-assisted analysis of MPs in environmental samples based on μFTIR imaging in combination with machine learning. *Environmental Science & Technology Letters*, *9*(1), 90–95. https://doi.org/10.1021/acs.estlett.1c00851.

Imavathy, S., & Chinnadurai, M. (2021). Threshold based support vector machine learning algorithm for sequential patterns. *International Journal of Computers Communications & Control*, *16*(6). https://doi.org/10.15837/ijccc.2021.6.4305.

Jie, Y. S., Wei, F. W., Zong-qi, C., & Qing, W. (2021). Study on rapid recognition of marine MPs based on raman spectroscopy. https://doi.org/10.3964/j.issn.1000-0593(2021)08-2469-05.

Jolliffe, I. T., & Cadima, J. (2016). Principal component analysis: A review and recent developments. *Philosophical Transactions of the Royal Society A: Mathematical, Physical and Engineering Sciences*, *374*(2065), 20150202. https://doi.org/10.1098/rsta.2015.0202.

Krogh, A., & Vedelsby, J. (1994). Neural network ensembles, cross validation, and active learning. *Advances in Neural Information Processing Systems*, *7*, 231–238.

Lenz, R., Enders, K., Stedmon, C. A., Mackenzie, D. M., & Nielsen, T. G. (2015). A critical assessment of visual identification of marine MP using Raman spectroscopy for analysis improvement. *Marine Pollution Bulletin*, *100*(1), 82–91. https://doi.org/10.1016/j.marpolbul.2015.09.026.

Li, L., Luo, Y., Peijnenburg, W. J., Li, R., Yang, J., & Zhou, Q. (2020). Confocal measurement of MPs uptake by plants. *MethodsX*, *7*, 100750.

Lorenzo-Navarro, J., Castrillón-Santana, M., Sánchez-Nielsen, E., Zarco, B., Herrera, A., Martínez, I., & Gómez, M. (2021). Deep learning approach for automatic MPs counting and classification. *Science of the Total Environment*, *765*, 142728.

Lusher, A. L., Burke, A., O'Connor, I., & Officer, R. (2014). MP pollution in the Northeast Atlantic Ocean: Validated and opportunistic sampling. *Marine Pollution Bulletin*, *88*(1–2), 325–333. https://doi.org/10.1016/j.marpolbul.2014.08.023.

Mansa, R., & Zou, S. (2021). Thermogravimetric analysis of MPs: A mini review. *Environmental Advances*, *5*, 100117. https://doi.org/10.1016/j.envadv.2021.100117.

Mariano, S., Tacconi, S., Fidaleo, M., Rossi, M., & Dini, L. (2021). Micro and nanoplastics identification: Classic methods and innovative detection techniques. *Frontiers in Toxicology*, *3*, 636640. https://doi.org/10.3389/ftox.2021.636640.

Martin, C., Parkes, S., Zhang, Q., Zhang, X., McCabe, M. F., & Duarte, C. M. (2018). Use of unmanned aerial vehicles for efficient beach litter monitoring. *Marine Pollution Bulletin*, *131*, 662–673.

Massarelli, C., Campanale, C., & Uricchio, V. F. (2021). A handy open-source application based on computer vision and machine learning algorithms to count and classify MPs. *Water*, *13*(15), 2104. https://doi.org/10.3390/w13152104.

Maxwell S, H., Melinda K, F., & Matthew, G. (2020). Counterstaining to separate Nile red-stained MP particles from terrestrial invertebrate biomass. *Environmental Science & Technology*, *54*(9), 5580–5588.

Melgani, F., & Bruzzone, L. (2004). Classification of hyperspectral remote sensing images with support vector machines. *IEEE Transactions on Geoscience and Remote Sensing*, *42*(8), 1778–1790. https://doi.org/10.1109/TGRS.2004.831865.

Meyers, N., Catarino, A. I., Declercq, A. M., Brenan, A., Devriese, L., Vandegehuchte, M., ... & Everaert, G. (2022). Microplastic detection and identification by Nile red staining: Towards a semi-automated, cost-and time-effective technique. *Science of the Total Environment*, *823*, 153441.

Monti, R. P., Tootoonian, S., & Cao, R. (2018). Avoiding degradation in deep feed-forward networks by phasing out skip-connections. In *Artificial Neural Networks and Machine Learning–ICANN 2018: 27th International Conference on Artificial Neural Networks, Rhodes, Greece, October 4–7, 2018, Proceedings, Part III 27* (pp. 447–456). Springer International Publishing. https://doi.org/10.1007/978-3-030-01424-7_44.

Prata, J. C., da Costa, J. P., Duarte, A. C., & Rocha-Santos, T. (2019). Methods for sampling and detection of MPs in water and sediment: A critical review. *TrAC Trends in Analytical Chemistry*, *110*, 150–159. https://doi.org/10.1016/j.trac.2018.10.029.

Primpke, S., Lorenz, C., Rascher-Friesenhausen, R., & Gerdts, G. (2017). An automated approach for MPs analysis using focal plane array (FPA) FTIR microscopy and image analysis. *Analytical Methods*, *9*(9), 1499–1511. https://doi.org/10.1039/C6AY02476A.

Shim, W. J., Hong, S. H., & Eo, S. (2017). Identification methods in MP analysis: A review. *Analytical Methods*, *9*, 1384–1391. https://doi.org/10.1039/C6AY02558G.

Shinzawa, H., Awa, K., Kanematsu, W., & Ozaki, Y. (2009). Multivariate data analysis for Raman spectroscopic imaging. *Journal of Raman Spectroscopy*, *40*(12), 1720–1725. https://doi.org/10.1002/jrs.2525.

Stapleton, P. A. (2021). MP and nanoplastic transfer, accumulation, and toxicity in humans. *Current Opinion in Toxicology*, *28*, 62–69. https://doi.org/10.1016/j.cotox.2021.10.001.

Stopar, K., & Bartol, T. (2019). Digital competences, computer skills and information literacy in secondary education: Mapping and visualization of trends and concepts. *Scientometrics*, *118*(2), 479–498.

Suárez-Araujo, C. P., García Báez, P., Sánchez Rodríguez, Á., & Santana-Rodrríguez, J. J. (2016). Supervised neural computing solutions for fluorescence identification of benzimidazole fungicides. Data and decision fusion strategies. *Environmental Science and Pollution Research*, *23*, 24547–24559. https://doi.org/10.1007/s11356-016-7129-8.

Wagner, M., Scherer, C., Alvarez-Muñoz, D., Brennholt, N., Bourrain, X., Buchinger, S., ... & Reifferscheid, G. (2014). MPs in freshwater ecosystems: What we know and what we need to know. *Environmental Sciences Europe*, *26*(1), 1–9. https://doi.org/10.1186/s12302-014-0012-7.

Wander, L., Vianello, A., Vollertsen, J., Westad, F., Braun, U., & Paul, A. (2020). Exploratory analysis of hyperspectral FTIR data obtained from environmental MPs samples. *Analytical Methods*, *12*(6), 781–791. https://doi.org/10.1039/C9AY02483B.

Wright, S. L., Levermore, J. M., & Kelly, F. J. (2019). Raman spectral imaging for the detection of inhalable MPs in ambient particulate matter samples. *Environmental Science & Technology*, *53*(15), 8947–8956. https://doi.org/10.1021/acs.est.8b06663.

Xu, J. L., Thomas, K. V., Luo, Z., & Gowen, A. A. (2019). FTIR and Raman imaging for MPs analysis: State of the art, challenges and prospects. *TrAC Trends in Analytical Chemistry*, *119*, 115629. https://doi.org/10.1016/j.trac.2019.115629.

Yamashita, R., Nishio, M., Do, R. K. G., & Togashi, K. (2018). Convolutional neural networks: An overview and application in radiology. *Insights into Imaging*, *9*, 611–629. https://doi.org/10.1007/s13244-018-0639-9.

Yu, F., & Hu, X. (2022). Machine learning may accelerate the recognition and control of MP pollution: Future prospects. *Journal of Hazardous Materials*, *432*, 128730. https://doi.org/10.1016/j.jhazmat.2022.128730.

Yurtsever, M., & Yurtsever, U. (2019). Use of a convolutional neural network for the classification of microbeads in urban wastewater. *Chemosphere*, *216*, 271–280.

Zhang, Y., Zhang, D., & Zhang, Z. (2023). A critical review on artificial intelligence—Based MPs imaging technology: Recent advances, hot-spots and challenges. *International Journal of Environmental Research and Public Health*, *20*(2), 1150. https://doi.org/10.3390/ijerph20021150.

Zhao, S., Qiu, Z., & He, Y. (2021). Transfer learning strategy for plastic pollution detection in soil: Calibration transfer from high-throughput HSI system to NIR sensor. *Chemosphere*, *272*, 129908. https://doi.org/10.1016/j.chemosphere.2021.129908.

Ziatdinov, M., Ghosh, A., Wong, C. Y., & Kalinin, S. V. (2022). AtomAI framework for deep learning analysis of image and spectroscopy data in electron and scanning probe microscopy. *Nature Machine Intelligence*, *4*(12), 1101–1112.

14 Artificial Intelligence and Machine Learning Approaches for Automatic Microplastics Identification and Characterization

Mohammad Karimi, Sorour Ayoubian Markazi, Mahdi Javanmardi, Masoumeh Sharifi Teshnizi, Haleh Khoramshahi, and Zohre Avakh

14.1 INTRODUCTION

The ubiquitous prevalence and long-term persistence of microplastics (MPs) in seas, rivers, lakes, and soils pose serious ecological concerns and endanger global environmental and biological diversity (Ayoubian Markazi et al., 2023; Xi et al., 2022). Accurate detection and analysis of MPs existence in the environment are essential to successfully combat their release from plastic debris and comprehend the possible threats they bring. This necessitates the development of precise detection algorithms capable of providing information on the size, number, and distribution of MPs. Furthermore, it's critical to pinpoint MP sources in order to create efficient mitigation techniques. This demands that different kinds of MP particles be chemically identified on an individual basis. Although it is a valid theory to anticipate MPs' fate and environmental effects based on their characteristics and complicated environmental circumstances, it is time-consuming, labor-intensive, and impracticable to undertake tests to confirm this idea. For dealing with such intricate interactions, conventional physical models fall short (Kannankai et al., 2022; Yu & Hu, 2022). Fortunately, the accelerated development of data analytics, artificial intelligence (AI), and machine learning (ML) techniques offers new opportunities for identifying MPs (Höppener et al., 2023). By combining ML with traditional MP identification methods, it is possible to overcome the limitations of conventional analytical techniques, resulting in enhanced qualitative and quantitative identification accuracy. This combination of conventional methods and machine learning produces a potent scientific tool for identifying and examining MP particulate matter in a wide array of environmental media (Lin et al., 2022).

DOI: 10.1201/9781003438793-14

Machine learning is related to a computer program that can be automatically improved with experience, i.e., characteristic parameters of microplastics. Many fields, including statistics, artificial intelligence, philosophy, information theory, computational complexity, and cognitive science, are involved in constructing the algorithms that are the core of machine learning. Several machine learning algorithms have been widely applied in the domain of microplastic pollution to overcome various issues and extract relevant insights from the data. Artificial neural networks (ANNs) are one technique that is frequently used. ANNs have been used for tasks such as the detection, categorization, and quantification of microplastics. Another common machine learning method in the field of microplastic research is support vector machines (SVMs). SVMs excel at binary classification and have been utilized to identify microplastics from other particles or debris. Random forest algorithms have also been effective in studying data on microplastics. Lastly, principal component analysis (PCA) has been applied to minimize the dimensionality of microplastic data while maintaining the most essential data. Other models, including decision trees (DT) and k-nearest neighbors (K-NNs), have also demonstrated promise in addressing various elements of microplastic pollution in addition to the machine learning techniques already discussed. Overall, these machine learning techniques have greatly advanced our knowledge of microplastic pollution and given us insightful information for environmental monitoring, evaluation, and management. Identification of MPs in the presence of other pollutants in wastewater, characterization of them at different WWT processes, and management of the whole system are difficult tasks because of huge amounts of information. Besides, traditional methods of identification and characterization do not have good enough precision, especially under the complexity of wastewater contaminants. Under this situation, ML algorithms provide efficient and quick answers to classify the MPs types, quantify the MPs concentration, and demonstrate their adsorption behavior.

All aforementioned algorithms which are core of machine learning techniques operate based on the human intelligence processes. Artificial Intelligence (AI) is the field of computing, analyzing, and inferring data using machines. Employing AI technology for MP removal is an important and most practical approach in the field of wastewater management. Rational agents are the center of AI. The nature of wastewater is a key factor affecting how well an agent can behave. An agent is a machine that receives information from the wastewater fluid through sensors and acts upon that wastewater treatment through actuators. The action is performed after any percepts; this connection from percepts to actions, named agent function, is mapped inside the AI by designing an agent program. All agents can improve their performance through learning. Learning involves acquiring general concepts from specific training examples including MPs and organic and inorganic matters. Learning improves the performance of agents, after observing the data of examples. When an agent is a computer, it is called machine learning. Indeed, a computer receives some data, creates a model based on the data, and practices the model that can identify or characterize the MPs. In this chapter, the methods of microplastics identification and characterization will be introduced and then most important machine learning algorithms trained for microplastics will be overviewed. Last part of this chapter includes an interesting approach related to the application of AI and ML in wastewater treatment and management. The efficiency and economics of microplastic removal at different processes of wastewater treatment plants (WWTPs) are the big challenge that can be solved by emerging technologies, that is, Fourth Industrial Revolution or Industry 4.0, integrating cloud computing, analytics, AI and machine learning into process facilities and throughout the operations.

14.2 METHODS OF MICROPLASTICS IDENTIFICATION AND CHARACTERIZATION

14.2.1 Vibrational Spectroscopy Methods

14.2.1.1 FTIR, ATR and Micro-FTIR

Fourier Transform Infrared (FTIR) spectroscopy facilitates the identification and characterization of the chemical bonds present in microplastics. This is achieved by measuring the absorption of infrared radiation by samples as a function of frequency or wavenumbers (in the range of 400–4,000 cm^{-1}) (Tirkey & Upadhyay, 2021). Next, the FTIR spectra obtained from the samples are compared with those of established plastic polymers in the spectral database, thus ensuring a high degree of accuracy (Veerasingam et al., 2021). Furthermore, it has the potential to provide additional insights into the physiochemical degradations of plastic particles by studying variations in absorption patterns resulting from the oxidation process (Shim et al., 2017). The technique of Attenuated Total Reflectance Fourier Transform Infrared Spectroscopy (ATR-FTIR) is frequently employed in the identification of microplastics exceeding the size range of 200–500 μm. Additionally, it is utilized for the analysis of weathered microplastics (MP) owing to its ability to provide valuable insights into the modified surfaces of the particles resulting from aging. This includes the presence of additional functional groups such as hydroxyl (OH), carbonyl (C=O), and carboxyl (COOH). Moreover, the application of ATR-FTIR is an effective method for distinguishing between synthetic and natural microfibers (Ivleva, 2021).

Micro-FTIR, a new development for the analysis of samples smaller than 20 μm, integrates FTIR spectroscopy with an IR microscope (Elkhatib & Oyanedel-Craver, 2020). Using focal plane array (FPA) detectors, the micro-FTIR approach enables chemical scanning of wider regions of membrane filters with excellent resolution. Although micro-FTIR imaging has proven to be a viable technique for the characterization of microplastics, it requires a significant amount of time for measurement, often exceeding 20 hours (Veerasingam et al., 2021).

14.2.1.2 Raman Spectroscopy

Raman spectroscopy is another vibration-based technique for detecting unidentified chemical compounds. This technique is a non-destructive method of spectroscopy that uses the inelastic scattering of photons from a monochromatic light source, typically a laser, to offer insights into the structural characteristics of polymers (Adhikari et al., 2022). In comparison with FTIR, Raman spectroscopy provides several advantages, including improved spatial resolution for small samples (<20 μm), higher sensitivity to non-polar functional groups, and narrow spectral bands (Tirkey & Upadhyay, 2021). In addition, generally, Raman vibration is observed to be strong in covalent bonds such as C–C, whereas FTIR vibration is observed to be strong in ionic bonds like O–H and N–H (Liu et al., 2022). Therefore, it is recommended that Raman spectroscopy and FTIR complement each other perfectly. On the other hand, while Raman spectroscopy has the potential to provide accurate spectra for microplastics with various shapes and dimensions, some interference caused by foreign bonds of additives results in limitations in analysis accuracy. In this case, the presence of dyes, pigments, and other organic compounds containing certain fluorescence groups should be considered. Therefore, the process of sample purification plays a crucial role in obtaining acceptable outcomes from the Raman analysis (Adhikari et al., 2022). Also, one potential approach for solving this limitation is to employ some automated algorithms to eliminate the fluorescence background from spectra. It is possible to correlate the polymer peaks with library matching software after background fluorescence subtraction to obtain information regarding microplastics chemical composition and other related data (Lei et al., 2022).

14.2.2 Visual Identification

Visual characterization methods are low-cost and practical traditional ways to minimize the number of particles that must be chemically identified subsequently (Nguyen et al., 2019). Based on the size, form, and color of the microplastics, visual techniques have been used to count and categorize them into different groups. Particles with a size greater than approximately 500 μm can be observed using either optical microscopy or visual observation (Adhikari et al., 2022). While it is possible to visually sort and identify large plastic particles that have distinct colors or morphologies, particles lacking identifiable shapes or colors pose a challenge for visual sorting without the aid of magnification tools. Electron microscopy, which offers enlarged images, serves a crucial role in the identification of microplastics that are difficult to discern (Fu et al., 2020).

The utilization of SEM in the detection of microplastics results in highly detailed and magnified images of polymer materials like microplastics through the emission of a high-intensity electron beam directed at the surface of the sample (Silva et al., 2018). The obtained clear images from particle surface make it possible to distinguish of microplastics from organic particles, especially the plastic-like ones. This technique is also employed to analyze the weathering and degradation of microplastics in their natural environment, specifically the changes in surface texture that result in the formation of pits and fractures (Liu et al., 2020). However, this microscopic technique has some limitations and disadvantages. Since microplastics are non-conductive, there is a need for sample preparation, including gold or carbon coating, which may hinder the morphological analysis of ultra-small microplastics (Adhikari et al., 2022). In addition, the procedure is high-cost and time-consuming and also poses a risk of erroneous results when comparing similar particles and not explaining the source or kind of polymer used as well as their color (Snega Priya et al., 2022).

The application of dyes for staining purposes is a useful technique in the identification of microplastics. The hydrophobic dye, Nile Red, is a commonly utilized dye that has demonstrated effectiveness in the detection of microplastics through fluorescence microscopy (Lv et al., 2021). Classification of microplastics into polar or hydrophobic categories is achieved depending on the polymer characteristics, which is based on the solvatochromic property of Nile Red that causes its color change in response to differences in solvent polarity (Meyers et al., 2022). Hence, it is feasible to differentiate polar polymers such as polyethylene terephthalate and polyamide from hydrophobic polymers, including polystyrene, polyethylene, and polypropylene (Silva et al., 2018). In this case, to overcome the challenges in distinguishing plastic and non-plastic particles, some machine learning-based automated and semi-automated methods are suggested (Meyers et al., 2022).

14.2.3 Thermo-Analytical Methods

Pyrolysis coupled with gas chromatography-mass spectrometry (py-GC-MS) is widely used for the identification of microplastics. This technique uses the thermal decomposition of microplastics under inert conditions to determine the chemical composition of the polymer through mass spectroscopy by comparing the spectra and reference data of known samples (Dümichen et al., 2017). However, since different polymers might generate similar products from pyrolysis, this technique runs the risk of misidentifying some types of microplastics. Additionally, the procedures used in GC-MS analysis are destructive, time-consuming, and operation-intensive. Furthermore, only a small quantity (0.5 mg) of samples could be evaluated, and it is not appropriate for analyzing heterogeneous and complicated samples (Zhu & Wang, 2020).

14.2.4 Automated Approaches

Numerous techniques are available for the identification and characterization of microplastics. Traditional spectroscopic and microscopic techniques were described earlier, nevertheless, there is still a lack of reliable detection of microplastics in different environments, especially wastewater plants (Jin et al., 2022). In some cases, microplastics are found in small amounts in the environment, which requires using highly sensitive detection methods. In addition, a significant problem in the study of environmental matrices is the existence of several kinds of natural particles and fibers in various quantities but in a similar size range as microplastics. Therefore, distinguishing between them through traditional methods becomes challenging (Liu et al., 2020). Moreover, the process of identifying and characterizing microplastics is laborious and time-consuming. To overcome the mentioned challenges, it is crucial to develop the traditional techniques of microplastic characterization (Kedzierski et al., 2019a). Recently, it is demonstrated that integrating machine learning techniques with described identification methodologies might greatly enhance the accuracy of analysis and classification of micro and nanoplastics. Furthermore, this combination will considerably increase the efficiency of detecting microplastics while also providing new theoretical and analytical guidance as well as a technological improvement foundation for additional MPs detection and identification (Lin et al., 2022).

14.3 OVERVIEW OF AI AND ML TECHNIQUES

In the past, the study of microplastics modeling faced challenges due to limited computational capabilities and a scarcity of extensive and readily available data. Computational methods were less capable, with restricted processing power and slower algorithms that hindered analysis of massive and complex datasets. Furthermore, there was a lack of data on microplastics, including information on their distribution, properties, and effects on the environment. The scarcity of data hampered the development of accurate and robust models (Zhang et al., 2023b). However, recent advances in computational power and the increasing availability of comprehensive datasets have revolutionized microplastic modeling. The capabilities of computational approaches have improved, making it possible to develop sophisticated machine learning algorithms and analyze large datasets efficiently. In addition, the availability of substantial and available data, ranging from field measurements to laboratory examinations, enables researchers to derive significant insights and make accurate predictions about microplastic behavior. These changes have improved the field of modeling microplastics, allowing for more precise assessments and intelligent decisions to solve the problems caused by microplastic pollution (Kannankai et al., 2022).

Machine learning is excellent at identifying significant correlations in highly complex and heterogeneous data, enabling quick prediction of unforeseen outcomes based on previously acquired knowledge (Zhang et al., 2023b). By enabling the prediction of environmental hazard outcomes and detecting crucial components that affect hazard risks, it has the potential to significantly speed up research development (Yu & Hu, 2022). The assessment of environmental pollution hazards has been made less complicated in the context of MPs by the effective application of ML techniques to object-oriented methods and remote sensing tools for quantifying and classifying plastic waste (Gonçalves et al., 2020). It is possible to improve the effectiveness and precision of MP identification in complex environments by combining ML techniques with traditional analysis techniques (Figure 14.1) (Astray et al., 2023; Lin et al., 2022). Not only do these innovations reduce the time and expense associated with analysis, but they also hold great promise for the field of MP identification (Lorenzo-Navarro et al., 2021).

14.3.1 Machine Learning Techniques

14.3.1.1 Decision Tree (DT) and Random Forest (RF)

The decision tree (DT) is a popular classification approach that creates a set of fundamental classifiers using training data. The base classifiers assist the ML model in estimating the value of a target variable by choosing the best-split attribute for classification at each internal node from the feature attributes of the sample data. However, when the number of samples in a dataset increases, DT can result in enormous, complicated trees that can lead to overfitting. By developing numerous separate decision trees and presenting their average prediction results, random forest (RF) significantly reduces the possibility of overfitting (Guo et al., 2021). In recent years, DT and RF have experienced substantial theoretical and methodological advancements.

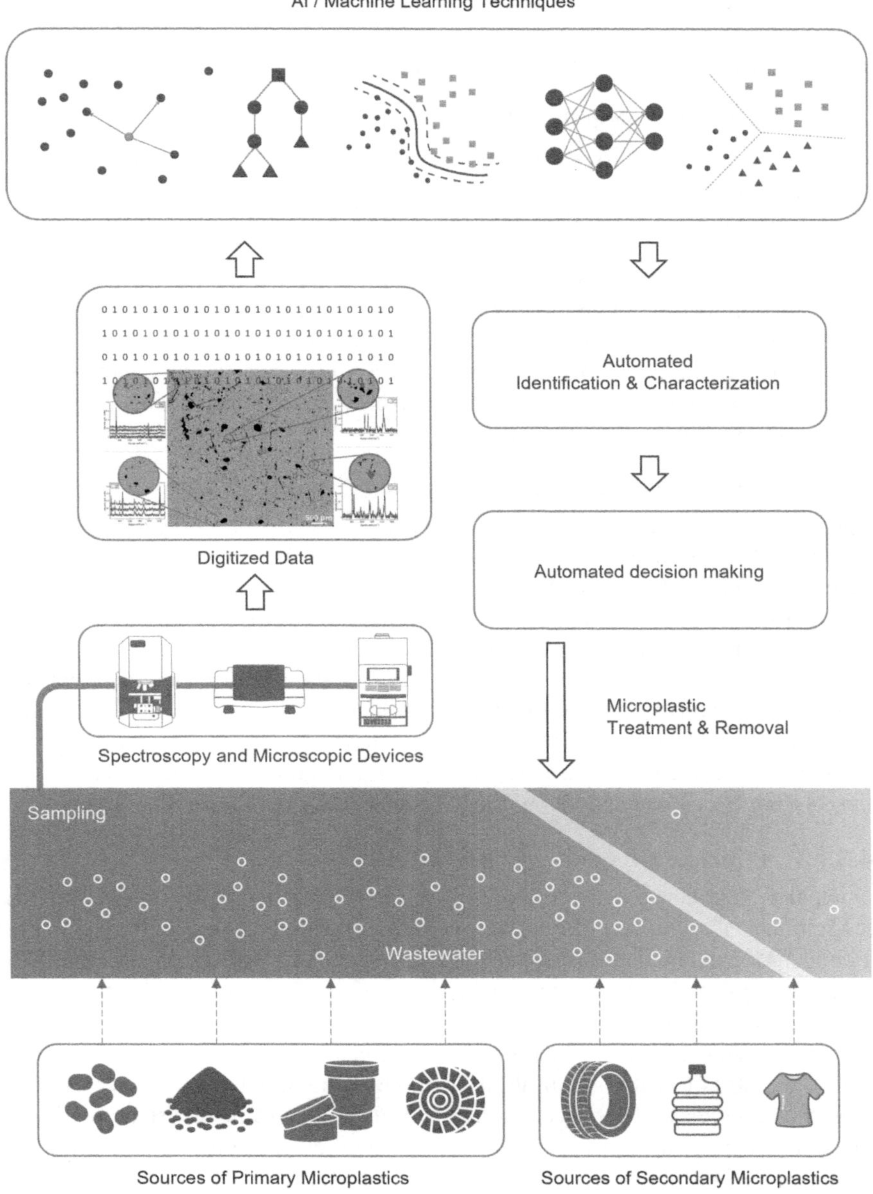

FIGURE 14.1 General architecture for automated identification and removal of the microplastics using AI/ML techniques.

As a result, they have gained significant popularity in ecological and environmental research. Furthermore, their application is gradually extending to the study of MPs in environmental media (Hufnagl et al., 2019).

14.3.1.2 K-Nearest Neighbor (k-NNs)

K-nearest neighbors (k-NNs) is a machine learning method that classifies or predicts the labels of data items based on their proximity. It is a non-parametric, supervised learning technique that selects the k points that are closest to an input and gives them the label that appears most frequently among them. K-NNs is a straightforward algorithm that can handle both categorical and numerical data (Tian et al., 2022). This algorithm can be effectively employed for the identification and classification of microplastics in environmental samples by utilizing Fourier Transform Infrared (FTIR) spectra. Based on spectral characteristics, we may distinguish between various microplastic polymer kinds, such as polyethylene, polypropylene, polystyrene, etc. by utilizing k-nearest (Yan et al., 2022).

14.3.1.3 Support Vector Machines (SVMs)

Support Vector Machines (SVMs) are supervised machine learning algorithms that are utilized for classification and regression applications. In a high-dimensional feature space, they seek to identify the best hyperplane to effectively divide various classes of data points (Imavathy & Chinnadurai, 2021; Shan et al., 2019). SVMs are useful for categorizing microplastic particles according to parameters such as size, shape, color, and spectral properties. They successfully discovered an ideal hyperplane that separates various microplastic classes by mapping the input data into a higher-dimensional feature space (Back et al., 2022). SVMs have been helpful in the study of microplastics in environmental media because they offer precise and reliable models for identifying microplastics over a wide variety of kinds and particle sizes (Lin et al., 2022).

14.3.1.4 Principal Component Analysis (PCA)

Principal Component Analysis (PCA) holds immense popularity and widespread recognition as a prominent multivariate machine learning technique. It aims to identify underlying characteristics from a multivariate dataset (in this case, spectral data) and is appropriate for exploratory analysis (unsupervised model) (Da Silva et al., 2020). By constructing a novel space comprising orthogonal variables, PCA effectively diminishes the dimensionality of data. This technique prioritizes the arrangement of variables based on their ability to explain the maximum variance observed in the original dataset. Feature extraction is frequently used in pattern recognition to reduce image noise using PCA and its updated method. The precision of MP identification could be improved by PCA's enhanced feature extraction capabilities (Luo et al., 2022a, 2022b; Wander et al., 2020).

14.3.1.5 Artificial Neural Networks (ANNs)

Artificial neural networks (ANNs) use a dynamic and adaptive network structure that simulates the connectivity of biological neurons. ANNs manage information progressively by distributing weights among their input, hidden, and output layers (Guo & Wang, 2021). These methods have become frequently employed in environmental research, especially when it comes to identifying and categorizing microplastics according to their spectroscopic or visual characteristics (Bhagat et al., 2020). Gathering images or spectroscopic data of microplastic samples, labeling them with relevant information, and using this labeled dataset to train the neural network are all parts of the training process. ANNs can successfully identify microplastics in novel, unseen samples using pattern recognition (Lin et al., 2022).

14.3.1.6 Deep Learning

Deep Learning, the most sophisticated type of machine learning, uses ANNs to solve computer vision issues such as image classification, object detection, and motion analysis (Shi et al., 2022). Deep Learning has achieved significant success in its pattern recognition applications, which makes it relevant for detecting MPs. Text, images, videos, and other types of data can all have their features extracted using a number of ANNs techniques (Lorenzo-Navarro et al., 2021). Deep neural networks (DNN), recurrent neural networks (RNN), and convolutional neural networks (CNN) are three examples of these techniques. Deep Neural Networks, which are used in deep learning, have several layers, or more than one layered neural network architecture. With more layers, the outcome will be more refined as each one gets closer to the final outcome (Michael Odhiambo et al., 2022).

14.3.1.7 Convolutional Neural Network (CNN)

A particular kind of artificial neural network which is frequently utilized for image recognition and analysis tasks is the convolutional neural networks (CNNs). The automatic feature extraction and learning capabilities of CNNs make them ideal for applications involving the identification and categorization of microplastics (Ng et al., 2020). The layers that make up CNNs include convolutional layers, pooling layers and fully connected layers. Convolutional layers extract features from input images by applying filters to detect edges, textures, and other visual patterns. The collected features' dimensionality is reduced by the pooling layers, which extract the most important data. Finally, the fully connected layers identify the features and predict them based on the learned patterns (Michael Odhiambo et al., 2022). On the other hand, vision transformers are a recent development in computer vision that has demonstrated promising outcomes for image recognition applications (Bazi et al., 2021). While their application in microplastics analysis is still in its early stages, vision transformers have the potential to complement or even exceed CNNs in certain contexts, providing an alternate approach for image recognition and analysis in microplastics research.

14.3.2 Microplastics Detection and Characterization Using AI and ML Techniques

The study of microplastics has undergone a revolutionary change in recent years thanks to the development of AI and ML techniques. Researchers have improved the identification, categorization, and comprehension of microplastics in a variety of environmental circumstances by utilizing the capabilities of computer approaches. Large datasets can be analyzed using AI and ML algorithms to reveal meaningful patterns and provide important insights into the properties and distribution of microplastics (Table 14.1) (Kannankai et al., 2022).

14.3.2.1 Image Processing and Analysis Techniques

Through the development of new techniques for defining the size and shape of microplastics using quantitative shape descriptors, image processing and analysis utilizing AI and ML can aid in the reduction of microplastics pollution. These techniques can enable an accurate characterization of the shape of microplastics by using widely accessible equipment for research and open-source software (Valente et al., 2023). Researchers can now thoroughly and impartially analyze hotspots and issues in the field of artificial intelligence-based microplastic imaging because of recent advancements in artificial intelligence technology. This entails looking into effective and affordable automatic quantification techniques for the shape, size, volume, and topological appearance characteristics of microplastics (Lorenzo-Navarro et al., 2021; Zhang et al., 2023b).

Massarelli et al. (2021) created a machine-learning-based method that can quickly and automatically count and categorize microplastics into four groups based on their shape and size without the need for manual steps. They developed an early machine learning method to count and categorize microplastics and supervised (k-nearest neighbors) and unsupervised classification techniques to evaluate the quantities of microplastic and characteristics. The size counting and classification method used by the machine learning algorithm showed promising outcomes, but it requires further improvements for visual classification. The supervised classification provided acceptable results with an accuracy that was consistently higher than 0.9. On the other hand, the unsupervised classification revealed a possible underestimation of some microplastic shape classifications as a result of the sampling technique used, leading to the development of a useful tool for introducing information that is indistinguishable by conventional research approaches used in microplastic studies. With significant potential for technique standardization, this application provides a reliable automated way for quantifying microplastics based on counts of particles captured in an image, size distribution, and morphology (Massarelli et al., 2021).

Shi et al. (2022) used SEM for examining microplastics from everyday items. They developed a deep learning technique to quantify and categorize these microplastics. The SEM method produced detailed images of microplastics, and deep learning models outperformed traditional computer vision techniques. The models precisely segmented microplastics and attained a high average Jaccard index. They also had a high accuracy of 98.33% when classifying microplastics based on their shapes. The trained models made it possible to quickly and accurately segment and classify new micrographs, replacing manual labor with an economical and effective method. The work shows that SEM and deep learning have a lot of potential for analyzing microplastics (Shi et al., 2022).

TABLE 14.1
AI and ML Applications for MPs Identification and Characterization

Purpose	Data Set	AI Method	References
Categorizing MPs in the 1–5 mm range	Pictures taken with a digital camera or a mobile phone	CNN (U-Net and VGG16)	Lorenzo-Navarro et al. (2021)
Quantification and shape classification of MPs	SEM images	CNN (U-Net and MultiResUNet)	Shi et al. (2022)
Automated microscale detection of MPs in a broad range within filtered water samples	Holo-graphic microscope	Holographic SVM	Bianco et al. (2020)
Identification and classification of the smallest MPs	SEM images with the spectroscopic power of cathodoluminescence (CL)	ANNs	Höppener et al. (2023)
Automated counting and classification of MPs into four morphology and size categories	A simple digital 12-megapixel smartphone camera	k-NNs	Massarelli et al. (2021)
Distinguishing between various polymer types, quantifying their abundances, and analyzing size distributions	μFTIR images	RF	Hufnagl et al. (2019)
Automated identification and classification of plastics	Raman Spectrum Matrix	PCA	Luo et al. (2022b)
Characterizing the spectral properties of MPs	Infrared spectra of various polymeric particles recorded by a quantum cascade laser (LDIR)	k-NNss, DT	Tian et al. (2022)
Identifying the spectra of classical polymers	FTIR spectra	k-NNs classification	Kedzierski et al. (2019b)
Predicting the relative abundance of Vibrio spp. on MPs	Adsorption capacity of MPs	SVM, deep learning, RF	Jiang et al. (2022)
Predicting the adsorption of organic contaminants onto MPs	Adsorption capacity of MPs	RF, SVM, ANNs	Astray et al. (2023)

Bianco et al. (2020) developed an automatic prescreening method that combines 3D coherent imaging with ML to detect MPs in pretreated water samples at a range of sizes. They discovered that MP recognition in heterogeneous samples is improved by ML based on an appropriate set of holographic features. They classified ten populations with over 99% accuracy, including an MP mixture made up of many different plastic materials and nine diatom species (DS). The residual classification errors during the test step were caused by misclassifications among diatom classes; no MPs were identified as a DS and vice versa. They were able to successfully identify MPs in pretreated seawater by using the suggested holographic SVMs as a binary classifier, which allowed them to ignore other objects with similar characteristic scale ranges (Bianco et al., 2020).

14.3.2.2 Spectroscopy Techniques and Chemical Analysis Techniques

Raman spectroscopy and FTIR have some serious drawbacks, such as poor resolution, long imaging times, and limited particle size analyses. Methods for ML have shown to be quite effective in increasing the accuracy of microplastics identification. Identification of microplastics could benefit from a combination of machine learning and traditional spectroscopy (Kannankai et al., 2022). MPs contamination can be effectively classified using FTIR spectroscopy in conjunction with the SVM regression approach. Back et al. found that SVM classifiers effectively overcome challenges such as large data volumes and poor spectral quality associated with FTIR imaging. This allows for rapid and automated characterization of microplastics in seawater (Back et al., 2022). In order to rapidly detect MPs in seawater, Lin et al. (2022) proposed an intelligent classification technique based on Raman spectral detection that integrates wavelet processing and the DT/RF algorithm. According to the results, RF can identify MPs with a cross-validation accuracy of up to 97.24%, which is higher than DT's cross-validation accuracy based on Raman spectroscopic data. This percentage can serve as a guideline for increasing the MP identification precision of Raman spectroscopy (Lin et al., 2022).

Researchers have shown that the power of PCA may be used to reduce the dimensionality of complicated datasets and reduce noise, improving the accuracy of MP recognition and allowing for the extraction of important data for MP identification (Luo et al., 2022b). The decoding of Raman spectral matrix imaging was carried out by Fang et al. (2022) using PCA, which successfully extracted and decoded important information from the spectrum matrix to improve the accuracy and interpretability of MPs samples (Fang et al., 2022). Wander et al. (2020) used PCA in combination with the FTIR approach to reduce the dimensionality of MP samples and demonstrate particle similarities (Wander et al., 2020).

To overcome the time-consuming and skill-dependent nature of analyzing spectral data, researchers conducted an automated learning method utilizing the KNN algorithm. Kedzierski et al. (2019b) carried out a study during the Tara Mediterranean Sea campaign, collecting over 4,000 FTIR spectra of microplastics. The method employed was effective in identifying specific polymers such as polyethylene, polypropylene, and polyamide. However, due to a lack of spectra in the learning database, it proved less effective for polymers like polyvinyl chloride. To increase the database's effectiveness, the researchers supported sharing FTIR spectrum databases from different studies and suggested adding more spectra to it (Kedzierski et al., 2019b).

14.3.2.3 Surface Properties Characterization

ML models excel at capturing complicated relationships and revealing non-linear patterns between input factors (e.g., microplastic properties, environmental circumstances) and output variables (e.g., surface attributes, adsorption behavior). A data-driven strategy enables ML algorithms to directly extract insightful information from enormous databases, producing precise predictions and a deeper comprehension of the behavior of microplastics. In addition, ML techniques excel at feature selection, adsorption process optimization, data adaptation, and spotting hidden patterns (Zhang et al., 2023a; Zhu et al., 2022b). The potential role of microplastics as a vector for the transmission of hazardous microorganisms, particularly *Vibrio* spp., which can cause serious foodborne illnesses in humans, was investigated using machine learning techniques. Jiang et al. (2022) developed machine learning models for predicting the relative abundance of Vibrio spp. on microplastics using datasets from an estuary and a mariculture zone in China. The outcomes showed that the DNN model and RT algorithm had the best-predicting performance. The prediction performance of the DNN and RT models was not significantly affected by alternative data sources, data sampling, or processing techniques (Jiang et al., 2022).

To better understand the interaction of microplastics with heavy metal ions, ANN models were developed based on literature data to predict the sorption capacity of heavy metal ions (including Cd, Pb, Cr, Cu, and Zn) onto microplastics in global aquatic environments. According to the findings, the ANN model could accurately estimate the sorption capacity. The amount anticipated and the field measurement results agreed, indicating that laboratory studies may be able to predict the capacity of heavy metal ions adsorption on microplastics in various aquatic environments (Guo & Wang, 2021). In conclusion, machine learning models can be effectively utilized in combination with spectral and visual analysis methods to enhance the analysis of microplastics. By integrating these techniques, a better understanding of microplastics' behavior can be achieved, leading to improved insights into their impact on the environment. This enhanced analysis enables the development of more effective strategies to control and mitigate the negative effects of microplastics on the environment.

14.4 MANAGEMENT AND ECONOMICS OF WWTPS

Numerous factors can affect the effluent quality of WWTPs, managing operations and maintenance of WWTPs can be difficult when expenses need to be managed. As a result, machine learning can be used more widely because it can give WWTP managers chances to lower expenses and enhance operations (Zhu et al., 2022a). Table 14.2 shows the AI and ML applications for the WWTPs with different algorithms. Evaluating inlet water quality plays a key role to set parameters for each section in WWTPs. Combining AI algorithms with soft sensors provides a tool that is more sensitive for real-time online monitoring. Several studies improved online monitoring of chlorine (Djerioui et al., 2018), ammonia (Cecconi & Rosso, 2021), *E. coli* (Foschi et al., 2021), COD and TSS (Qin et al., 2011). Devices can collect and send data thanks to a large-scale technology known as the Internet of Things (IoT). In addition, combining IoT and AI algorithms to predict and model water quality measurements can help with monitoring, decision-making, improving the quality of the water and automating process. The energy used for wastewater treatment is significantly influenced by the pump system. The fluid flow rate and pump energy consumption were modeled using the ANNs to reduce energy consumption with a model error of under 3% (Zhang et al., 2012). In pipes, flow-measurement sensors are typically applied. But measurement inaccuracy driven on by contaminants, corrosion, and extreme turbidity might result in incorrect readings. By improving existing sensors, deep learning has the ability to increase measurement accuracy in a variety of circumstances (Ji et al., 2020).

TABLE 14.2
AI and ML Applications for the WWTPs and Common Algorithms

Process	Algorithm	Input	Output	References
Coagulation and flocculation process	ANNs, MLP, Elman neural network (ENN), adaptive neuro fuzzy inference system (ANFIS)	Dissolved oxygen, oxygen consumption, temperature, water flow, pH, settling time, and raw water turbidity	Turbidity removal, flocculant dosage, coagulant dosage	Safeer et al. (2022)
Biological process	ANNs, SVM, CNN, Gaussian process regression (GPR), PCA	TSS, HRT, harvesting period time, sodium acetate addition, light intensity, temperature, pH, and NO^{-3} concentration	Estimating the growth of alga mixed culture, prediction of total nitrogen	Sundui et al. (2021)
Advanced oxidation process	ANNs, Gradient Boosting Machine (GBM), gradient boosting decision trees (GBDT)	Catalyst dose, pollutant concentration, pH, UV light regions, reusability of the catalyst	Degradation efficiency	Hassani et al. (2015), Navidpour et al. (2022), Jaffari et al. (2023)
Membrane filtration process	ANNs, fuzzy logic (FL), genetic programming (GP), genetic algorithm (GA), particle swarm optimization (PSO)	Feed water, turbidity, transmembrane pressure, Filtration time, concentration, and pH	Specific cake resistance, permeate flux	Bagheri et al. (2019)
Management and economics of WWTPs	Neural network (NN),	Pump speeds, level of the raw wastewater junction	Energy consumption of pump system	Zhang et al. (2012)
	Deep learning and image process	Image of inside the pipe	Wastewater level in pipe and surface flow velocity measurements	Ji et al. (2020)

AI and ML algorithms have demonstrated the capability to identify microplastics in wastewater. By incorporating detection algorithms for microplastics in WWTPs, it is possible to optimize various parameters automatedly in different processes and efficiently remove microplastics at a low cost with minimal operational challenges.

14.4.1 Limitations and Challenges of Current AI and ML Approaches in MPs Characterization and Identification

Microplastic characterization and identification utilizing ML and AI have demonstrated significant potential in recent years, but several limitations exist. One important challenge is related to the diversity and complexity of microplastics. ML models are trained on particular datasets, and their accuracy can be compromised when faced with variations in microplastic types, sizes, shapes, and conditions. As a result, these algorithms can be unable to accurately identify less common or irregular microplastics, potentially leading to misclassification in real-world environmental samples (Zhang et al., 2023b). Additionally, the need for extensive data preprocessing and quality control is another limitation. Microplastic identification and characterization involve complex imaging or spectroscopic techniques, and the data provided can be noisy or incomplete. For precise identification, ML and AI algorithms need clean, reliable data; therefore, quality control and data preprocessing are important but time-consuming stages in the process (Luo et al., 2022b).

Another major limitation is the scarcity of publicly available data. Since microplastic research requires specialized and costly equipment, public data sources are limited, and scientists often have to generate their own datasets (Kannankai et al., 2022). Therefore, ML and AI algorithms developed on these particular datasets may have limited publicity and applicability beyond the original research context, preventing broader adoption of these technologies (Yu & Hu, 2022; Zhang et al., 2023a). Furthermore, a lack of standardization in data collecting and analysis techniques among research groups and equipment can hinder the development of generally applicable ML and AI models. Variations in sampling methods, data preprocessing, and instrumentation may result in inconsistencies in datasets, making it difficult to produce models that generalize well across different environments and conditions (Lin et al., 2022; Yan et al., 2022).

Despite these drawbacks, ML and AI have a lot of potential to improve our knowledge of microplastics and promote the development of solutions that mitigate their negative effects on the environment. In order to overcome these challenges and utilize ML and AI in the fight against microplastic contamination, further research and collaboration are required.

14.5 CONCLUSION AND PROSPECTIVE

In conclusion, the study of microplastics has been revolutionized by the application of AI and ML approaches. More precise identification, characterization, and modeling of microplastics are now possible thanks to improved computational capabilities and the availability of large and available datasets. AI and ML have substantially improved our understanding of microplastic contamination by analyzing complex datasets, revealing hidden patterns, and making accurate predictions about microplastic behavior and effects. These methods have improved the efficiency of microplastic identification, enabling researchers to examine a vast number of samples and classify microplastics into different categories according to their size, shape, material, and surface characteristics. Furthermore, the predictive models developed using AI and ML allow for the forecasting of microplastic movement, fate, and ecological interactions, hence facilitating informed decision-making and effective mitigation methods. Looking forward, the future of microplastics research using AI and ML holds immense potential. With the advent of Industry 4.0, AI and ML approaches can be utilized to optimize sorting and recycling procedures, assisting in the effective separation of microplastics and fostering sustainable resource management. Additionally, AI and ML can be applied in WWTPs for real-time monitoring and management of microplastics, improving their removal efficiency and reducing environmental contamination. As the area advances, interdisciplinary collaborations and developments in AI and ML will drive innovation in microplastic research, enabling comprehensive solutions to reduce the effects of microplastic pollution and protect the environment and human well-being.

REFERENCES

Adhikari, S., Kelkar, V., Kumar, R., & Halden, R. U. (2022). Methods and challenges in the detection of microplastics and nanoplastics: A mini-review. *Polymer International*, *71*(5), 543–551. https://doi.org/10.1002/pi.6348.

Astray, G., Soria-Lopez, A., Barreiro, E., Mejuto, J. C., & Cid-Samamed, A. (2023). Machine learning to predict the adsorption capacity of microplastics. *Nanomaterials*, *13*(6), 1061. https://doi.org/10.3390/nano13061061.

Ayoubian Markazi, S., Karimi, M., Yousefi, B., Sadati, M., Khoramshahi, H., Khoee, S., & Karimi, M. R. (2023). Experimental and modeling study on the simultaneous fouling behavior of micro/nanoplastics and bovine serum albumin in ultrafiltration membrane separation. *Journal of Environmental Chemical Engineering*, *11*(2). https://doi.org/10.1016/j.jece.2023.109354.

Back, H. de M., Vargas Junior, E. C., Alarcon, O. E., & Pottmaier, D. (2022). Training and evaluating machine learning algorithms for ocean microplastics classification through vibrational spectroscopy. *Chemosphere*, *287*. https://doi.org/10.1016/j.chemosphere.2021.131903.

Bagheri, M., Akbari, A., & Mirbagheri, S. A. (2019). Advanced control of membrane fouling in filtration systems using artificial intelligence and machine learning techniques: A critical review. *Process Safety and Environmental Protection*, *123*, 229–252. https://doi.org/10.1016/j.psep.2019.01.013.

Bazi, Y., Bashmal, L., Al Rahhal, M. M., Dayil, R. Al, & Ajlan, N. Al. (2021). Vision transformers for remote sensing image classification. *Remote Sensing*, *13*(3), 1–20. https://doi.org/10.3390/rs13030516.

Bhagat, S. K., Tung, T. M., & Yaseen, Z. M. (2020). Development of artificial intelligence for modeling wastewater heavy metal removal: State of the art, application assessment and possible future research. *Journal of Cleaner Production*, *250*. Elsevier Ltd. https://doi.org/10.1016/j.jclepro.2019.119473.

Bianco, V., Memmolo, P., Carcagnì, P., Merola, F., Paturzo, M., Distante, C., & Ferraro, P. (2020). Microplastic identification via holographic imaging and machine learning. *Advanced Intelligent Systems*, *2*(2), 1900153. https://doi.org/10.1002/aisy.201900153.

Cecconi, F., & Rosso, D. (2021). Soft sensing for on-line fault detection of ammonium sensors in water resource recovery facilities. *Environmental Science Technology*, *55*, 14, 10067–10076. https://doi.org/10.1021/acs.est.0c06111.

Da Silva, V. H., Murphy, F., Amigo, J. M., Stedmon, C., & Strand, J. (2020). Classification and quantification of microplastics (<100 μm) using a focal plane array-Fourier transform infrared imaging system and machine learning. *Analytical Chemistry*, *92*(20), 13724–13733. https://doi.org/10.1021/acs.analchem.0c01324.

Djerioui, M., Bouamar, M., Ladjal, M., & Zerguine, A. (2018). Chlorine soft sensor based on extreme learning machine for water. *Arabian Journal for Science and Engineering*, *Cl*. https://doi.org/10.1007/s13369-018-3253-8.

Dümichen, E., Eisentraut, P., Gerhard, C., Barthel, A., Senz, R., & Braun, U. (2017). Chemosphere Fast identi fi cation of microplastics in complex environmental samples by a thermal degradation method. *Chemosphere*, *174*, 572–584. https://doi.org/10.1016/j.chemosphere.2017.02.010.

Elkhatib, D., & Oyanedel-Craver, V. (2020). A critical review of extraction and identification methods of microplastics in wastewater and drinking water. *Environmental Science and Technology*, *54*(12), 7037–7049. https://doi.org/10.1021/acs.est.9b06672.

Fang, C., Luo, Y., Zhang, X., Zhang, H., Nolan, A., & Naidu, R. (2022). Identification and visualisation of microplastics via PCA to decode Raman spectrum matrix towards imaging. *Chemosphere*, *286*. https://doi.org/10.1016/j.chemosphere.2021.131736.

Foschi, J., Turolla, A., & Antonelli, M. (2021). Soft sensor predictor of *E. coli* concentration based on conventional monitoring parameters for wastewater disinfection control. *Water Research*, *191*. https://doi.org/10.1016/j.watres.2021.116806.

Fu, W., Min, J., Jiang, W., Li, Y., & Zhang, W. (2020). Separation, characterization and identification of microplastics and nanoplastics in the environment. *Science of the Total Environment*, *721*, 137561. https://doi.org/10.1016/j.scitotenv.2020.137561.

Gonçalves, G., Andriolo, U., Gonçalves, L., Sobral, P., & Bessa, F. (2020). Quantifying marine macro litter abundance on a sandy beach using unmanned aerial systems and object-oriented machine learning methods. *Remote Sensing*, *12*(16). https://doi.org/10.3390/RS12162599.

Guo, H., Wu, S., Tian, Y., Zhang, J., & Liu, H. (2021). Application of machine learning methods for the prediction of organic solid waste treatment and recycling processes: A review. In *Bioresource Technology* (Vol. 319). Elsevier Ltd. https://doi.org/10.1016/j.biortech.2020.124114.

Guo, X., & Wang, J. (2021). Projecting the sorption capacity of heavy metal ions onto microplastics in global aquatic environments using artificial neural networks. *Journal of Hazardous Materials*, *402*. https://doi.org/10.1016/j.jhazmat.2020.123709.

Hassani, A., Khataee, A., & Karaca, S. (2015). Photocatalytic degradation of ciprofloxacin by synthesized TiO_2 nanoparticles on montmorillonite: Effect of operation parameters and artificial neural network modeling. *Journal of Molecular Catalysis A: Chemical*, *409*, 149–161. https://doi.org/10.1016/j.molcata.2015.08.020.

Höppener, E. M., Shahmohammadi, M. (Sadegh), Parker, L. A., Henke, S., & Urbanus, J. H. (2023). Classification of (micro)plastics using cathodoluminescence and machine learning. *Talanta*, *253*. https://doi.org/10.1016/j.talanta.2022.123985.

Hufnagl, B., Steiner, D., Renner, E., Löder, M. G. J., Laforsch, C., & Lohninger, H. (2019). A methodology for the fast identification and monitoring of microplastics in environmental samples using random decision forest classifiers. *Analytical Methods*, *11*(17), 2277–2285. https://doi.org/10.1039/c9ay00252a.

Imavathy, S., & Chinnadurai, M. (2021). Threshold based support vector machine learning algorithm for sequential patterns. *International Journal of Computers, Communications and Control*, *16*(6). https://doi.org/10.15837/IJCCC.2021.6.4305.

Ivleva, N. P. (2021). *Chemical Analysis of Microplastics and Nanoplastics: Challenges, Advanced Methods, and Perspectives*. https://doi.org/10.1021/acs.chemrev.1c00178.

Jaffari, Z. H., Abbas, A., Lam, S., Park, S., Chon, K., Kim, E., & Cho, K. H. (2023). Machine learning approaches to predict the photocatalytic performance of bismuth ferrite-based materials in the removal of malachite green. *Journal of Hazardous Materials*, *442*(September 2022), 130031. https://doi.org/10.1016/j.jhazmat.2022.130031.

Ji, H. W., Yoo, S. S., Lee, B. J., Koo, D. D., & Kang, J. H. (2020). Measurement of wastewater discharge in sewer pipes using image analysis. *Water (Switzerland)*, *12*(6), 1–10. https://doi.org/10.3390/w12061771.

Jiang, J., Zhou, H., Zhang, T., Yao, C., Du, D., Zhao, L., Cai, W., Che, L., Cao, Z., & Wu, X. E. (2022). Machine learning to predict dynamic changes of pathogenic Vibrio spp. abundance on microplastics in marine environment. *Environmental Pollution*, *305*. https://doi.org/10.1016/j.envpol.2022.119257.

Jin, M., Liu, J., Yu, J., Zhou, Q., Wu, W., Fu, L., Yin, C., Fernandez, C., & Karimi-Maleh, H. (2022). Current development and future challenges in microplastic detection techniques: A bibliometrics-based analysis and review. *Science Progress*, *105*(4), 1–22. https://doi.org/10.1177/00368504221132151.

Kannankai, M. P., Babu, A. J., Radhakrishnan, A., Alex, R. K., Borah, A., & Devipriya, S. P. (2022). Machine learning aided meta-analysis of microplastic polymer composition in global marine environment. *Journal of Hazardous Materials*, *440*. https://doi.org/10.1016/j.jhazmat.2022.129801.

Kedzierski, M., Falcou-Préfol, M., Kerros, M. E., Henry, M., Pedrotti, M. L., & Bruzaud, S. (2019a). A machine learning algorithm for high throughput identification of FTIR spectra: Application on microplastics collected in the Mediterranean Sea. *Chemosphere*, *234*, 242–251. https://doi.org/10.1016/j.chemosphere.2019.05.113.

Lei, B., Bissonnette, J. R., Hogan, Ú. E., Bec, A. E., Feng, X., & Smith, R. D. L. (2022). Customizable machine-learning models for rapid microplastic identification using Raman microscopy. *Analytical Chemistry*, *94*(49), 17011–17019. https://doi.org/10.1021/acs.analchem.2c02451

Lin, J. U, Liu, H. T, & Zhang, J. (2022). Recent advances in the application of machine learning methods to improve identification of the microplastics in environment. *Chemosphere*, *307*(P4), 136092. https://doi.org/10.1016/j.chemosphere.2022.136092.

Liu, M., Lu, S., Chen, Y., Cao, C., Bigalke, M., & He, D. (2020). Analytical methods for microplastics in environments: Current advances and challenges. In *Handbook of Environmental Chemistry* (Vol. 95, pp. 3–24). Springer. https://doi.org/10.1007/698_2019_436.

Liu, Y., Wang, B., Pileggi, V., & Chang, S. (2022). Methods to recover and characterize microplastics in wastewater treatment plants. *Case Studies in Chemical and Environmental Engineering*, *5*(January), 100183. https://doi.org/10.1016/j.cscee.2022.100183.

Lorenzo-Navarro, J., Castrillón-Santana, M., Sánchez-Nielsen, E., Zarco, B., Herrera, A., Martínez, I., & Gómez, M. (2021). Deep learning approach for automatic microplastics counting and classification. *Science of the Total Environment*, *765*. https://doi.org/10.1016/j.scitotenv.2020.142728.

Luo, Y., Su, W., Xu, X., Xu, D., Wang, Z., Wu, H., Chen, B., & Wu, J. (2022a). Raman spectroscopy and machine learning for microplastics identification and classification in water environments. *IEEE Journal of Selected Topics in Quantum Electronics*, *29*(4), 6900308. https://doi.org/10.1109/JSTQE.

Luo, Y., Zhang, X., Zhang, Z., Naidu, R., & Fang, C. (2022b). Dual-principal component analysis of the Raman spectrum matrix to automatically identify and visualize microplastics and nanoplastics. *Analytical Chemistry*, *94*(7), 3150–3157. https://doi.org/10.1021/acs.analchem.1c04498.

Lv, L., Yan, X., Feng, L., Jiang, S., Lu, Z., Xie, H., Sun, S., Chen, J., & Li, C. (2021). Challenge for the detection of microplastics in the environment. *Water Environment Research*, *93*(1), 5–15. https://doi.org/10.1002/wer.1281.

Massarelli, C., Campanale, C., & Uricchio, V. F. (2021). A handy open-source application based on computer vision and machine learning algorithms to count and classify microplastics. *Water (Switzerland)*, *13*(15). https://doi.org/10.3390/w13152104.

Meyers, N., Catarino, A. I., Declercq, A. M., Brenan, A., Devriese, L., Vandegehuchte, M., De Witte, B., Janssen, C., & Everaert, G. (2022). Microplastic detection and identification by Nile red staining: Towards a semi-automated, cost- and time-effective technique. *Science of the Total Environment*, *823*, 153441. https://doi.org/10.1016/j.scitotenv.2022.153441.

Michael Odhiambo, J., Mvurya, M., Luvanda, A., Mwakondo, F., & Michael Odhiambo, J. (2022). Deep learning algorithm for identifying microplastics in open sewer systems: A systematic review. *The International Journal of Engineering and Science (IJES)*, *11*(5), 11–18. https://doi.org/10.9790/1813-1105011118.

Navidpour, A. H., Hosseinzadeh, A., Huang, Z., Li, D., & Zhou, J. L. (2022). Application of machine learning algorithms in predicting the photocatalytic degradation of perfluorooctanoic acid. *Catalysis Reviews*, *62*(2), 1–26. https://doi.org/10.1080/01614940.2022.2082650.

Nguyen, B., Claveau-Mallet, D., Hernandez, L. M., Xu, E. G., Farner, J. M., & Tufenkji, N. (2019). Separation and analysis of microplastics and nanoplastics in complex environmental samples. *Accounts of Chemical Research*, *52*(4), 858–866. https://doi.org/10.1021/acs.accounts.8b00602.

Qin, X., Gao, F., & Chen, G. (2011). Wastewater quality monitoring system using sensor fusion and machine learning techniques. *Water Research*, *46*(4), 1133–1144. https://doi.org/10.1016/j.watres.2011.12.005.

Safeer, S., Pandey, R. P., Rehman, B., Safdar, T., Ahmad, I., Hasan, S. W., & Ullah, A. (2022). A review of artificial intelligence in water purification and wastewater treatment: Recent advancements. *Journal of Water Process Engineering*, *49*(June), 102974. https://doi.org/10.1016/j.jwpe.2022.102974.

Shan, J., Zhao, J., Zhang, Y., Liu, L., Wu, F., & Wang, X. (2019). Simple and rapid detection of microplastics in seawater using hyperspectral imaging technology. *Analytica Chimica Acta*, *1050*, 161–168. https://doi.org/10.1016/j.aca.2018.11.008.

Shi, B., Patel, M., Yu, D., Yan, J., Li, Z., Petriw, D., Pruyn, T., Smyth, K., Passeport, E., Miller, R. J. D., & Howe, J. Y. (2022). Automatic quantification and classification of microplastics in scanning electron micrographs via deep learning. *Science of the Total Environment*, *825*. https://doi.org/10.1016/j.scitotenv.2022.153903.

Shim, W. J., Hong, S. H., & Eo, S. E. (2017). Identification methods in microplastic analysis: A review. *Analytical Methods*, *9*(9), 1384–1391. https://doi.org/10.1039/c6ay02558g.

Silva, A. B., Bastos, A. S., Justino, C. I. L., da Costa, J. P., Duarte, A. C., & Rocha-Santos, T. A. P. (2018). Microplastics in the environment: Challenges in analytical chemistry - A review. *Analytica Chimica Acta*, *1017*, 1–19. https://doi.org/10.1016/j.aca.2018.02.043.

Snega Priya, P., Kamaraj, M., Aravind, J., & Muthukumaran, P. (2022). Microplastics sampling and recovery: Materials, identification, characterization methods and challenges. In *Environmental Footprints and Eco-Design of Products and Processes* (pp. 155–175). Springer, Singapore. https://doi.org/10.1007/978-981-16-8440-1_8.

Sundui, B., Ramirez Calderon, O. A., Abdeldayem, O. M., Lázaro-Gil, J., Rene, E. R., & Sambuu, U. (2021). Applications of machine learning algorithms for biological wastewater treatment: Updates and perspectives. *Clean Technologies and Environmental Policy*, *23*(1), 127–143. https://doi.org/10.1007/s10098-020-01993-x.

Tian, X., Beén, F., & Bäuerlein, P. S. (2022). Quantum cascade laser imaging (LDIR) and machine learning for the identification of environmentally exposed microplastics and polymers. *Environmental Research*, *212*. https://doi.org/10.1016/j.envres.2022.113569.

Tirkey, A., & Upadhyay, L. S. B. (2021). Microplastics: An overview on separation, identification and characterization of microplastics. *Marine Pollution Bulletin*, *170*(June), 112604. https://doi.org/10.1016/j.marpolbul.2021.112604.

Valente, T., Ventura, D., Matiddi, M., Sbrana, A., Silvestri, C., Piermarini, R., Jacomini, C., & Costantini, M. L. (2023). Image processing tools in the study of environmental contamination by microplastics: Reliability and perspectives. *Environmental Science and Pollution Research, 30*(1), 298–309. https://doi.org/10.1007/s11356-022-22128-3.

Veerasingam, S., Ranjani, M., Venkatachalapathy, R., Bagaev, A., Mukhanov, V., Litvinyuk, D., Mugilarasan, M., Gurumoorthi, K., Guganathan, L., Aboobacker, V. M., & Vethamony, P. (2021). Contributions of Fourier transform infrared spectroscopy in microplastic pollution research: A review. *Critical Reviews in Environmental Science and Technology, 51*(22), 2681–2743. https://doi.org/10.1080/10643389.2020.1807450.

Wander, L., Vianello, A., Vollertsen, J., Westad, F., Braun, U., & Paul, A. (2020). Exploratory analysis of hyperspectral FTIR data obtained from environmental microplastics samples. *Analytical Methods, 12*(6), 781–791. https://doi.org/10.1039/c9ay02483b.

Xi, B., Wang, B., Chen, M., Lee, X., Zhang, X., Wang, S., Yu, Z., & Wu, P. (2022). Environmental behaviors and degradation methods of microplastics in different environmental media. *Chemosphere, 299*. https://doi.org/10.1016/j.chemosphere.2022.134354.

Yan, X., Cao, Z., Murphy, A., & Qiao, Y. (2022). An ensemble machine learning method for microplastics identification with FTIR spectrum. *Journal of Environmental Chemical Engineering, 10*(4). https://doi.org/10.1016/j.jece.2022.108130.

Yu, F., & Hu, X. (2022). Machine learning may accelerate the recognition and control of microplastic pollution: Future prospects. *Journal of Hazardous Materials, 432*. https://doi.org/10.1016/j.jhazmat.2022.128730.

Zhang, W., Huang, W., Tan, J., Huang, D., Ma, J., & Wu, B. (2023a). Modeling, optimization and understanding of adsorption process for pollutant removal via machine learning: Recent progress and future perspectives. In *Chemosphere* (Vol. 311). Elsevier Ltd. https://doi.org/10.1016/j.chemosphere.2022.137044.

Zhang, Y., Zhang, D., & Zhang, Z. (2023b). A critical review on artificial intelligence—based microplastics imaging technology: Recent advances, hot-spots and challenges. *International Journal of Environmental Research and Public Health, 20*(2). https://doi.org/10.3390/ijerph20021150.

Zhang, Z., Zeng, Y., & Kusiak, A. (2012). Minimizing pump energy in a wastewater processing plant. *Energy, 47*(1), 505–514. https://doi.org/10.1016/j.energy.2012.08.048.

Zhu, J., & Wang, C. (2020). Recent advances in the analysis methodologies for microplastics in aquatic organisms: Current knowledge and research challenges. *Analytical Methods, 12*(23), 2944–2957. https://doi.org/10.1039/d0ay00143k.

Zhu, M., Wang, J., Yang, X., Zhang, Y., Zhang, L., Ren, H., Wu, B., & Ye, L. (2022a). A review of the application of machine learning in water quality evaluation. *Eco-Environment & Health, 1*(2), 107–116. https://doi.org/10.1016/j.eehl.2022.06.001.

Zhu, T., Tao, C., Cheng, H., & Cong, H. (2022b). Versatile in silico modelling of microplastics adsorption capacity in aqueous environment based on molecular descriptor and machine learning. *Science of the Total Environment, 846*. https://doi.org/10.1016/j.scitotenv.2022.157455.

Index

For Product Safety Concerns and Information please contact our EU representative GPSR@taylorandfrancis.com Taylor & Francis Verlag GmbH, Kaufingerstraße 24, 80331 München, Germany

Batch number: 10397790

Printed by Printforce, the Netherlands